High Dive

Circular Functions and the Physics of Falling Objects

Teacher's Guide

This material is based upon work supported by the National Science Foundation under award numbers ESI-9255262, ESI-0137805, and ESI-0627821. Any opinions, findings, and conclusions or recommendations expressed in this publication are those of the authors and do not necessarily reflect the views of the National Science Foundation.

Key Curriculum
1150 65th Street
Emeryville, California 94608
email: editorial@keypress.com
www.keycurriculum.com

First Edition Authors

Dan Fendel, Diane Resek, Lynne Alper, and Sherry Fraser

Contributors to the Second Edition

Sherry Fraser, Jean Klanica, Brian Lawler, Eric Robinson, Lew Romagnano, Rick Marks, Dan Brutlag, Alan Olds, Mike Bryant, Jeri P. Philbrick, Lori Green, Matt Bremer, Margaret DeArmond

Editor

Josephine Noah

Editorial Assistant

Emily Reed

Professional Reviewer

Rick Marks, Sonoma State University

Math Checker

Carrie Gongaware

Production Director

Christine Osborne

Production Editor

Andrew Jones

Executive Editor

Josephine Noah

Mathematics Product Manager

Elizabeth DeCarli

Publisher

Steven Rasmussen

Contents

Blackline Masters

Calculator Guide and **Calculator Notes**

High Dive Overview

Intent

In this unit, students study trigonometry in the context of a unit problem that involves a circus act.

Mathematics

Here is a summary of the main concepts and skills that students will encounter and practice in this unit:

Trigonometry

- Extending the trigonometric functions to all angles
- Reinforcing the importance of similarity in the definitions of the trigonometric functions
- Graphing the trigonometric functions and variations on those functions
- Defining the inverse trigonometric functions and principal values
- Discovering and explaining the Pythagorean identity $\sin^2 \theta + \cos^2 \theta = 1$, and other trigonometric identities
- Defining polar coordinates and finding rectangular coordinates from polar coordinates and vice versa

Physics

- Developing quadratic expressions for the height of free-falling objects, based on the principle of constant acceleration
- Recognizing that a person falling from a moving object will follow a different path than someone falling from a stationary object

Quadratic Equations

- Developing simple quadratic equations to describe the behavior of falling objects

Other concepts and skills are developed in connection with Problems of the Week.

Progression

The central problem of this unit concerns a circus act in which a diver is dropped from a turning Ferris wheel into a tub of water carried by a moving cart. The basic problem is to determine when his fall should begin in order for him to land in the water. Students begin by looking at the diver's height off the ground while still on the Ferris wheel, analyzing different cases in terms of the angle through which the wheel has turned, and seeing that the analysis is slightly different from one quadrant to another.

Students then develop a formula based on right-triangle trigonometry that works when the diver is in the first quadrant. They use this formula as a clue for how to extend the sine function from the familiar right-triangle context so that it is defined for all angles. The development of the general definition of the sine function involves several considerations, including the physical situation of the Ferris wheel, a graph of the diver's height while on the wheel, and a coordinate model of the wheel.

The cosine function is developed similarly and reinforces the elegance and power of definitions involving the coordinate system. In particular, students see that these definitions eliminate the need for a quadrant-by-quadrant analysis, incorporating the issue of sign quite nicely.

Students also learn about the graphs of the sine and cosine functions. In particular, they see how the graphs of the functions describing the diver's position change in response to various parameters such as the radius of the Ferris wheel and the period of its motion. As students search for angles corresponding to given values of the sine and cosine, they develop the concept of inverse trigonometric functions and their principal values.

The physics and mathematics of falling objects compose another major strand of the unit, based on the principle that falling objects have constant acceleration. Students are told to assume for simplicity that at the instant the dive begins the diver is falling as if from rest, even though this contradicts the physics of the actual situation. (The realistic solution of the circus dive, taking into account the diver's initial velocity, is found in the Year 4 unit *The Diver Returns*.) Students develop an expression for the height of an object falling from rest in terms of its time in the air, which is then used to determine the duration of the diver's fall. In the process, they review ideas about instantaneous and average speed and interpret these concepts graphically.

This work on falling objects is then combined with the analysis of the diver's position on the Ferris wheel and with information about the speed of the cart. Students synthesize these parts of the problem to develop a complex equation involving the amount of time the diver should stay on the Ferris wheel before being dropped. Then they solve this equation graphically (because it is too complex to be susceptible to algebraic manipulation) to find a solution to the unit problem, subject to the simplifying assumption described above.

After solving the unit problem, students pursue other aspects of trigonometric functions, including polar coordinates, a Pythagorean identity, and other trigonometric identities.

The overall organization of the unit can be summarized as follows:

Going to the Circus: Introducing the unit problem

The Height and the Sine: Analyzing the diver's height while still on the wheel and developing the sine function

Falling, Falling, Falling: Studying constant acceleration and developing a formula for the height of an object falling from rest

Moving Left and Right: Analyzing the diver's horizontal position and developing the cosine function

Finding the Release Time: Solving the unit problem, subject to the simplifying assumption that the moving diver falls as if from rest

A Trigonometric Conclusion: Continuing to work with trigonometric functions, including polar coordinates and identities

Pacing Guides

50-Minute Pacing Guide (23 days)

Day	Activity	Time Estimate
1	*Going to the Circus*	0
	The Circus Act	30
	Introduce: *POW 14: The Tower of Hanoi*	15
	Homework: *The Ferris Wheel*	5
2	Discussion: *The Ferris Wheel*	15
	Discussion: *The Circus Act*	35
	Homework: *As the Ferris Wheel Turns*	0
3	Discussion: *As the Ferris Wheel Turns*	20
	The Height and the Sine	0
	At Certain Points in Time	30
	Homework: *A Clear View*	0
4	Discussion: *A Clear View*	10
	Extending the Sine	20
	Testing the Definition	20
	Homework: *Graphing the Ferris Wheel*	0
5	Discussion: *Testing the Definition*	35
	Discussion: *Graphing the Ferris Wheel*	15
	Homework: *Ferris Wheel Graph Variations*	0
6	Discussion: *Ferris Wheel Graph Variations*	15
	The "Plain" Sine Graph	35
	Homework: *Sand Castles*	0
7	Discussion: *Sand Castles*	25
	Presentations: *POW 14: The Tower of Hanoi*	20
	Introduce: *POW 15: Paving Patterns*	5
	Homework: *More Beach Adventures*	0
8	Discussion: *More Beach Adventures*	10
	Falling, Falling, Falling	0
	Distance with Changing Speed	40
	Homework: *Acceleration Variations and a Sine Summary*	0
9	Discussion: *Acceleration Variations and a Sine*	10

	Summary	
	Free Fall	40
	Homework: *Not So Spectacular*	0
10	Discussion: *Not So Spectacular*	25
	Discussion: *Free Fall*	25
	Homework: *A Practice Jump*	0
11	Discussion: *A Practice Jump*	25
	Moving Left and Right	0
	Cart Travel Time	25
	Homework: *Where Does He Land?*	0
12	Discussion: *Where Does He Land?*	20
	First Quadrant Platform	30
	Homework: *Carts and Periodic Problems*	0
13	Discussion: *Carts and Periodic Problems*	10
	Generalizing the Platform	40
	Homework: *Planning for Formulas*	0
14	Discussion: *Planning for Formulas*	30
	Finding the Release Time	0
	Moving Cart, Turning Ferris Wheel (continue tomorrow)	20
	Homework: *Putting the Cart Before the Ferris Wheel*	0
15	Discussion: *Putting the Cart Before the Ferris Wheel*	10
	Moving Cart, Turning Ferris Wheel (continued)	40
	Homework: *What's Your Cosine?*	0
16	Discussion: *What's Your Cosine?*	20
	Discussion: *Moving Cart, Turning Ferris Wheel*	30
	Homework: *Find the Ferris Wheel*	0
17	Discussion: *Find the Ferris Wheel*	10
	A Trigonometric Conclusion	0
	Some Polar Practice	40
	A Polar Summary	0
	Homework: *Polar Coordinates on the Ferris Wheel*	0
18	Discussion: *Polar Coordinates on the Ferris Wheel*	10
	Pythagorean Trigonometry	40

	Homework: *Coordinate Tangents*	0
19	Discussion: *Coordinate Tangents*	20
	Positions on the Ferris Wheel	30
	Homework: *More Positions on the Ferris Wheel*	0
20	Discussion: *More Positions on the Ferris Wheel*	15
	Presentations: *POW 15: Paving Patterns*	35
	Homework: *A Trigonometric Reflection*	0
21	Discussion: *A Trigonometric Reflection*	20
	Speculation on the more complex version of the unit problem	30
	Homework: *"High Dive" Portfolio*	0
22	In-Class Assessment	40
	Homework: *Take-Home Assessment*	10
23	Exam Discussion	40
	Unit Reflection	10

50-Minute Pacing Guide (16 days)

Day	Activity	Time Estimate
1	*Going to the Circus*	0
	The Circus Act	70
	Introduce: *POW 14: The Tower of Hanoi*	15
	Homework: *The Ferris Wheel*	5
2	Discussion: *The Ferris Wheel*	10
	As the Ferris Wheel Turns	50
	The Height and the Sine	0
	At Certain Points in Time	30
	Homework: *A Clear View*	0
3	Discussion: *A Clear View*	10
	Extending the Sine	25
	Testing the Definition	55
	Homework: *Graphing the Ferris Wheel*	0
4	Discussion: *Graphing the Ferris Wheel*	15
	Ferris Wheel Graph Variations	40
	The "Plain" Sine Graph	35
	Homework: *Sand Castles*	0
5	Discussion: *Sand Castles*	30
	More Beach Adventures	50
	Falling, Falling, Falling	0
	Homework: *Distance with Changing Speed*	10
6	Discussion: *Distance with Changing Speed*	20
	Presentations: *POW 14: The Tower of Hanoi*	25
	Introduce: *POW 15: Paving Patterns*	5
	Free Fall	40
	Homework: *Acceleration Variations and a Sine Summary*	0
7	Discussion: *Acceleration Variations and a Sine Summary*	10
	Discussion: *Free Fall*	25
	Not So Spectacular	55

	Homework: *A Practice Jump*	0
8	Discussion: *A Practice Jump*	15
	Moving Left and Right	0
	Cart Travel Time	30
	Where Does He Land?	45
	Homework: *First Quadrant Platform*	0
9	Discussion: *First Quadrant Platform*	5
	Carts and Periodic Problems	40
	Generalizing the Platform	45
	Homework: *Planning for Formulas*	0
10	Discussion: *Planning for Formulas*	40
	Finding the Release Time	0
	Putting the Cart Before the Ferris Wheel	50
	Homework: *What's Your Cosine?*	0
11	Discussion: *What's Your Cosine?*	15
	Presentations: *POW 15: Paving Patterns*	25
	Moving Cart, Turning Ferris Wheel	50
	Homework: *Find the Ferris Wheel*	0
12	Discussion: *Find the Ferris Wheel*	10
	Discussion: *Moving Cart, Turning Ferris Wheel*	30
	A Trigonometric Conclusion	0
	Some Polar Practice	50
	A Polar Summary	0
	Homework: *Polar Coordinates on the Ferris Wheel*	0
13	Discussion: *Polar Coordinates on the Ferris Wheel*	10
	Pythagorean Trigonometry	40
	Coordinate Tangents	40
	Homework: *Positions on the Ferris Wheel*	0
14	Discussion: *Positions on the Ferris Wheel*	5
	Discussion: *Coordinate Tangents*	15
	More Positions on the Ferris Wheel	40
	A Trigonometric Reflection	30
	Homework: *"High Dive" Portfolio*	0
15	Discussion: *A Trigonometric Reflection*	20
	In-Class Assessment	40

	Homework: *Take-Home Assessment*	30
16	Exam Discussion	50
	Unit Reflection	10
	Speculation on the more complex version of the unit problem	30

Materials and Supplies

All IMP classrooms should have a set of standard supplies, described in the section "Materials and Supplies for the IMP Classroom" in *A Guide to IMP.* You'll also find a comprehensive list of materials needed for all Year 3 units in the section "Materials and Supplies for Year 3" in the *Year 3 Teacher's Guide* general resources.

No additional supplies are needed for this unit. However, general and activity-specific blackline masters, for transparencies or for student worksheets, are available in the "Blackline Masters" section in *High Dive* Unit Resources.

High Dive *Materials*

- Items for building models of the problem, such as paper plates, pipe cleaners, and toy cars
- Coins or discs

More About Supplies

Graph paper is a standard supply for IMP classrooms. Blackline masters of 1-Centimeter Graph Paper, ¼-Inch Graph Paper, and 1-Inch Graph Paper are provided, for you to make copies and transparencies.

Assessing Progress

High Dive concludes with two formal unit assessments. In addition, there are many opportunities for more informal, ongoing assessments throughout the unit. For more information about assessment and grading, including general information about the end-of-unit assessments and how to use them, consult *A Guide to IMP*.

End-of-Unit Assessments

This unit concludes with in-class and take-home assessments. The in-class assessment is intentionally short so that time pressures will not affect student performance. Students may use graphing calculators and their notes from previous work when they take the assessments. You can download unit assessments from the *High Dive* Unit Resources.

Ongoing Assessment

One of the primary tasks of the classroom teacher is to assess student learning. Although the assigning of course grades may be part of this process, assessment more broadly includes the daily work of determining how well students understand key ideas and what level of achievement they have attained on key skills, in order to provide the best possible ongoing instructional program for them.

Students' written and oral work provides many opportunities for teachers to gather this information. We make some recommendations here of written assignments and oral presentations to monitor especially carefully that will give you insight into student progress.

- *As the Ferris Wheel Turns*
- *Testing the Definition*
- *More Beach Adventures*
- *A Practice Jump*
- *Moving Cart, Turning Ferris Wheel*

Discussion of Unit Assessments

Have students volunteer to explain their work on each of the problems. Encourage questions and alternate explanations from other students.

In-Class Assessment

You might have the presenter begin by sketching a graph of the height of Walter's back. The choice of what point in the cycle to use for $t = 0$ is arbitrary, and students' graphs will vary depending on this choice. Using $t = 0$ to represent a time when Walter is at the midpoint in his cycle, on his way up (when his back is 10 feet below the surface), the graph of one complete cycle would look like this:

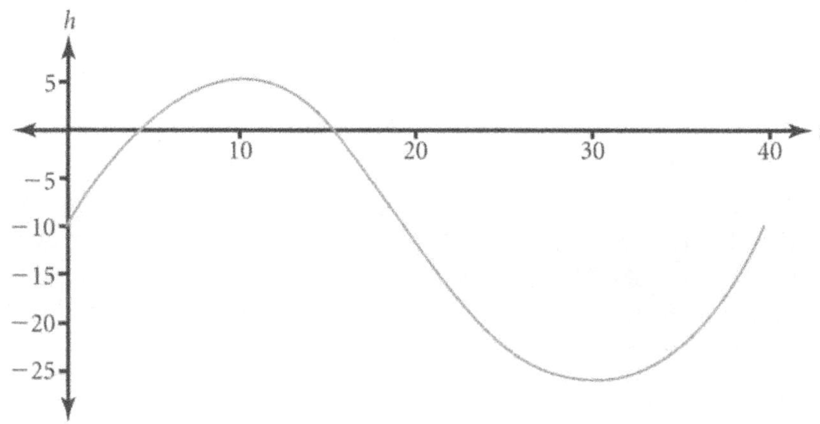

In terms of this graph, students need to know what fraction of the curve is above the horizontal axis. They might do this by solving the equation $15 \sin \theta = 10$, which gives the approximate solutions $\theta = 42°$ and $\theta = 138°$. The difference, 96°, is roughly 27 percent of 360°, so the probability that a person will see Walter immediately is approximately .27.

Take-Home Assessment

Question 1 is an application of two ideas of the unit: polar coordinates and movement of a free-falling object. Students will likely first calculate the height of the pebble when it releases using the equation $\sin 125 = \frac{y}{80}$, finding that y is approximately 65.5 ft. Students may then use the fact that when an object falls from rest, it takes $\sqrt{\frac{h}{16}}$ seconds to fall h feet. So, the pebble will reach the ground in about 2.02 seconds.

On Question 2, students should see that for first-quadrant angles, sec θ is equal to the ratio $\frac{r}{x}$, and conclude that this ratio should be used for all angles. They may bring out that in right triangles, sec θ is equal to $\frac{1}{\cos\theta}$, and use that relationship as an additional justification for the general definition of the secant function. (If students do not note that sec θ is equal to $\frac{1}{\cos\theta}$, you should mention this yourself.)

You may want to take a minute to discuss the graph of the secant function. Bring out that the function is undefined for θ = 90° and 270° and that these are the values for which the cosine function is 0. This diagram shows how the graph should look:

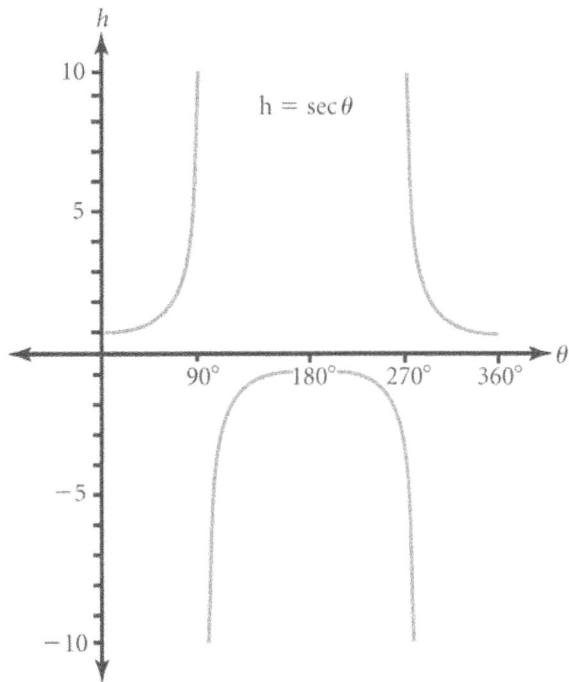

Extending cotangent and cosecant

To complete the work of extending the trigonometric functions, ask students to name the trigonometric functions that they have not yet extended. Review the fact that there are six ratios possible using the lengths of the sides of a right triangle (perhaps reviewing the notation ${}_3P_2$ at the same time).

Review the right-triangle definitions for the other two functions, cotangent and cosecant, and the fact that each is the reciprocal of one of the "basic" functions, tangent and sine. Also take a few minutes to look at the graph of each of these functions, bringing out that the graph of the cotangent function is a shifted and reversed version of the graph of the tangent function, and that the graph of the cosecant function is a shifted version of the graph of the secant function.

Going to the Circus

Intent

In this section, students are introduced to the central unit problem and begin to explore the relationship between the angular speed of the Ferris wheel and the height of the platform.

Mathematics

The first element of the unit problem that students will wrestle with is finding the height of the Ferris wheel's diving platform at any given time. Students begin by using right-triangle trigonometry to find the height of various clock positions on the Ferris wheel. Then the motion of the wheel is introduced, and the students use the period of revolution to find the angular velocity and thus the angular position and height of the platform at a given time.

Progression

The Circus Act introduces the central unit problem. In *The Ferris Wheel*, students find the height of the platform at given clock positions. In *As the Ferris Wheel Turns*, they work with the relationship between elapsed time and the height of the platform on the moving Ferris wheel.

The Circus Act

POW 14: The Tower of Hanoi

The Ferris Wheel

As the Ferris Wheel Turns

The Circus Act

Intent

This activity introduces the central unit problem.

Mathematics

As students explore the central unit problem, they must decide how the information they are given is relevant to the solution as they construct physical models of the situation. The activity also requires that they analyze what other information is needed that has not been provided.

Progression

The Circus Act asks students to make a model of the Ferris wheel situation that is described and to make a list of other information that will be needed in order to solve the unit problem. This is primarily an opportunity to ensure that students thoroughly understand the situation and question of the unit problem.

The discussion following the activity reveals some further basic facts about the Ferris wheel setup.

Approximate Time

30 minutes for activity
35 to 40 minutes for discussion

Classroom Organization

Small groups, followed by whole-class discussion

Materials

Items for building models of the problem, such as paper plates, pipe cleaners, and toy cars
Transparency or poster of *The Circus Act* blackline master

Doing the Activity

You may want to have one or more students read the introduction to the unit problem out loud. Then provide materials to each group so that students can make their own physical model of the problem (as in Question 1). You might provide paper plates and pipe cleaners for the Ferris wheel itself and a toy car to represent the moving cart. Circulate among the groups to see that everyone has the right idea. (You may want to keep one or more of these models on hand for demonstration purposes during the rest of the unit.)

The cart and Ferris wheel are set up similarly to what is shown in the diagram below, with the cart passing in front of the Ferris wheel. The platform, which is not shown in this diagram, is pointed "forward" (out from the page), perpendicular to the plane in which the Ferris wheel turns. The front end of the platform is directly above the path of the cart.

Once everyone has a clear understanding of the situation, each group should compile a list of questions as indicated in Question 2.

Discussing and Debriefing the Activity

Have groups share their lists of questions from Question 2. You may want to distinguish, as described here, between questions about the Ferris wheel setup and more general questions about falling objects. As the questions are proposed, you or a student can record them on chart paper.

Here are some of the questions students might ask about the setup:

- What is the radius of the Ferris wheel?
- How high is the center of the Ferris wheel above the ground?
- How fast does the Ferris wheel turn around?
- In what direction (clockwise or counterclockwise) does the Ferris wheel turn?
- When the cart starts moving, what is its position in relation to the Ferris wheel? (That is, how far is the cart from the Ferris wheel, and in what direction?)
- How fast does the cart go?
- How high is the water level in the cart above the ground?
- Where is the diver in the Ferris wheel's cycle when the cart starts moving?

Facts About the Ferris Wheel

After students have posed their questions, present them with the following information about the circus Ferris wheel and the high-dive act. Give them all of this information even if they didn't specifically ask for it. Tell them that the description of this Ferris wheel will be used throughout the unit, though occasionally they will consider other Ferris wheels too (such as in *The Ferris Wheel*).

You may want to make a poster or transparency of the diagram of the Ferris wheel in *The Circus Act* blackline master) and mark in each detail as it is discussed—the diagram does not show the numerical details:

- The Ferris wheel has a radius of 50 feet.
- The center of the Ferris wheel is 65 feet off the ground.
- The Ferris wheel turns at a constant speed, making a complete turn every 40 seconds.
- The Ferris wheel turns counterclockwise.
- When the cart starts moving, its center is 240 feet to the left of the center of the base of the Ferris wheel.
- The cart travels to the right at a constant speed of 15 feet per second.
- The water level in the cart is 8 feet above the ground.
- When the cart starts moving, the diver's platform is at the 3 o'clock position in its cycle.

Students should assume that when the cart starts moving, it is immediately going at 15 feet per second.

This picture shows what the final diagram of the situation might look like:

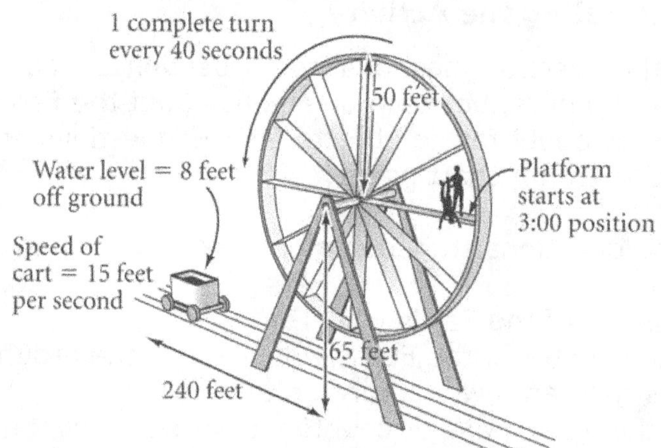

Portions of diagram not to scale.

Post all of the Ferris wheel and cart information prominently in the classroom, together with the diagram. Some of this information will be used right away (in *As the Ferris Wheel Turns*) while other information will not be needed for a while.

Note: This is the first of many posters that you will make for this unit. Be sure to have plenty of wall space (or arrange a way to place posters in a flip-chart arrangement, which takes up less space). Also, encourage students to take notes on these facts and subsequent formulas, because they will need this information.

Students may have a variety of questions about the physics of falling objects. That is, they may wonder exactly what happens to the diver once he is released. Here are some questions that may arise:

- How fast does the diver fall?
- How long does it take the diver to reach the ground?
- Do these answers depend on the diver's weight? On his height?

Tell students that they will be learning about the mathematics and physics of falling objects later in the unit, but that they will not need to answer questions like these just yet.

For Teachers: The Diver's Initial Motion

In this unit, we will be simplifying the problem by assuming that once the diver is released, he falls straight down as if he had fallen from a motionless Ferris wheel. In *Moving Cart, Turning Ferris Wheel*, students will solve the problem based on that assumption.

In the Year 4 unit *The Diver Returns,* students will return to this situation and deal with the fact that once the diver is released, his path depends not only on where he is released but also on the direction and magnitude of the initial speed he gets from the motion of the turning Ferris wheel.

If this complication comes up in today's discussion, you should acknowledge that the diver does not fall straight down as if from rest. Tell students that they will eventually take this into account. For now, however, they will be dealing with a simplified version of the problem in which he does fall straight down as if from rest.

Supplemental Activity

Mr. Ferris and His Wheel (extension) provides interested students with an opportunity to research the general topic of Ferris wheels.

POW 14: The Tower of Hanoi

Intent

In this activity, students solve a classic mathematical puzzle involving recursion.

Mathematics

The Tower of Hanoi is a classic puzzle that challenges students to find the smallest number of moves in which 64 stacked graduated disks can be moved from one of three posts to another, one at a time, while never placing a larger disk on top of a smaller one. Students are asked for two solutions, one that is based on a recursive process and another that uses a closed formula. Students explain generalizations, which essentially requires a proof.

The discussion of the POW solution introduces recursion and inductive reasoning.

Progression

This is a difficult problem, and students should be given a week or more to work on it. As usual, several student presentations are scheduled for this POW.

Approximate Time

15 minutes for introduction
1 to 3 hours for activity (at home)
20 to 25 minutes for presentations and discussion

Classroom Organization

Individuals, followed by several student presentations and whole-class discussion

Materials

Coins or discs

Doing the Activity

Have students act out a simple version of the puzzle described in *POW 14: The Tower of Hanoi* to be sure they understand the rules. You can use the case of two discs to focus on the number of moves required. Students should see that the pile cannot be moved in fewer than three moves. You may want to let students work on the case of three discs in groups today.

Be sure students notice that the write-up categories for this POW are somewhat different from the standard categories.

Give students about a week to work on this activity. On the day before the POW is due, choose three students to make POW presentations on the following day, and give them overhead transparencies and pens to take home to use for preparing those presentations.

Discussing and Debriefing the Activity

Have the three students chosen give their presentations, and then let other students share their ideas. Students should at least find the number of moves required for specific small numbers of discs.

Finding the Minimum Number

You can use the cases involving only a few discs to bring up the issue of how to show that these results are the smallest possible number of moves. For instance, by carefully examining the case of three discs, students should note the principle that it's "wasteful" to move the same disc on two consecutive moves. This principle shows that in the first three moves, one should get the two smallest discs onto another peg and that this can't be done in fewer than three moves. You can work further with this case to see why it's impossible to move all three discs in fewer than seven moves.

Students might use the information about specific cases in various ways to develop a general formula. If no general formulas are presented, you might have students put the results from individual cases into a table like this:

Number of discs	Number of moves needed
1	1
2	3
3	7
4	15

Ask, **What patterns or rules did you find?** Two main observations are likely to come out of this table and out of students' work in generating it. If we use a_n to represent the minimum number of moves required if there are n discs, then the two key patterns can be written as:

$$a_n = 2^n - 1$$
$$a_{n+1} = 2a_n + 1$$

One proof of the first relationship involves the second relationship, together with an approach similar to mathematical induction. Use your judgment about how much of the discussion presented here will be appropriate for your students.

The Closed Form: $a_n = 2^n - 1$

Some of your students likely will see that the entries are each one less than a power of 2. In other words, the general formula for the number of moves required to move *n* discs is $2^n - 1$.

The Recursive Pattern: $a_{n+1} = 2a_n + 1$

Another pattern is that each successive entry is obtained by doubling the previous entry and adding 1. Ask, **Why does this pattern hold true?** Some students may have observed this pattern and seen why it works in the course of doing examples. If so, they can probably articulate fairly well why it works. If not, one way to help students get some insight into this is to have a student act out the process for, say, four discs, but to interrupt after three discs have been moved to another peg. Help them to see the process of moving four discs in three stages:

- Move three discs to the middle peg (7 moves)
- Move the biggest disc to the right peg (1 move)
- Move three discs from the middle peg to the right (7 moves)

They should notice that they cannot move the fourth disc until the other three are all on the same one of the other two pegs.

It is important to articulate that moving three discs from one peg to another is the same no matter which two pegs are involved. That's why the first and last stages in the three-stage process just described each take the same number of moves as solving the original puzzle for three discs. You simply have to be careful about which peg you move the first disc to.

To reinforce this idea, you might ask what the largest number of discs was for which anybody actually carried out the process. Suppose, for example, someone did eight discs, in 255 moves. Ask, **How can you use the answer for eight discs to help you get the answer for nine discs?** Students should see that they can move eight of the discs to the middle in 255 moves, use one move to move the largest disc to the right, and then use 255 more moves to move the pile of eight from the middle to the right. This gives a total of 255 + 1 + 255 = 511 moves.

In other words, if we use a_n to represent the number of moves required to move an *n*-disc pile, then we see that this relationship holds:

$$a_{n+1} = a_n + 1 + a_n$$

(Students may simplify this to $a_{n+1} = 2a_n + 1$, but you might prefer to use the unsimplified formula because that reflects the sequence in which the groups of moves are done.)

Introduce the word *recursion* to describe the process by which each term in a sequence is described in terms of a preceding term or terms. The relationship $a_{n+1} = a_n + 1 + a_n$ (or $a_{n+1} = 2a_n + 1$) is called the *recursive formula.*

Proving the Closed Formula

If you have discussed both the recursive formula and the *closed formula* $a_n = 2^n - 1$, ask, **Why does the closed formula work?** Point out that they can see that it works by example for the specific cases in the table, but that doesn't mean that it always works.

You can ask students to imagine that they know that the rule $2^n - 1$ works for a specific large value of n, such as $n = 20$. That is, they should suppose that they have verified somehow that moving 20 discs requires $2^{20} - 1$ moves. Ask, **Can you find the number of moves for 21 discs without multiplying out 2^{20}?** Tell them that they can use the recursive pattern. They should see that 21 discs will require $(2^{20} - 1) + 1 + (2^{20} - 1)$ moves.

Ask, **What do you get when you simplify $(2^{20} - 1) + 1 + (2^{20} - 1)$?** They should see that the expression simplifies to $2 \cdot 2^{20} - 1$, which is the same as $2^{21} - 1$.

Ask, **Will this work for any number of discs?** Students should see that it does, and you can ask them to try to produce a general argument that if n discs can be transferred in $2^n - 1$ moves, then $n + 1$ discs can be transferred in $2^{n+1} - 1$ moves. As discussed earlier, the three-stage process gives the formula

$$a_{n+1} = a_n + 1 + a_n$$

so if n discs take $2^n - 1$ moves, then $n + 1$ discs take $(2^n - 1) + 1 + (2^n - 1)$ moves. But this simplifies to $2^{n+1} - 1$ moves.

In other words, this reasoning shows that if the formula $2^n - 1$ works for a particular number of discs, it also works when there is one more disc. If students get this far, tell them that this explanation is an example of a form of proof called *mathematical induction.* Tell them that this type of proof involves two elements: getting started (such as showing that the formula works in the case of only one disc) and going from one stage to the next (such as proving the recursive formula).

What About the Monks?

Oh yes, let's not forget the monks. With 64 discs, the task would take $2^{64} - 1$ moves, or approximately $1.8 \cdot 10^{19}$. A series of computations shows that $1.8 \cdot 10^{19}$ seconds is about 580 billion years. Doing only 40 discs would take a mere 35,000 years. (Siddhartha Gautama, founder of Buddhism, lived from about

563 B.C.E. to 483 B.C.E. If the monks had begun this task when he was alive and moved one disc per second, they would now have moved the thirty-seventh disc and would be rebuilding the pile of the first 36 discs on top of it.)

Key Questions

What patterns or rules did you find?

Why does this pattern hold true?

How can you use the answer for eight discs to help you get the answer for nine discs?

Why does the closed formula work?

Can you find the number of moves for 21 discs without multiplying out 2^{20}?

What do you get when you simplify $(2^{20} - 1) + 1 + (2^{20} - 1)$?

Will this work for any number of discs?

The Ferris Wheel

Intent

In this activity, students recognize that trigonometry will play a role in the unit problem.

Mathematics

This activity gives students a chance to investigate a Ferris wheel's motion in a more elementary context than that of the main problem. They begin to use trigonometry to calculate the height of various points on the wheel.

Progression

Students work individually to find the height at several clock positions, then discuss their results as a class.

Approximate Time

5 minutes for introduction
25 minutes for activity (at home or in class)
10 to 15 minutes for discussion

Classroom Organization

Individuals, followed by whole-class discussion

Materials

Optional: Transparency of *The Ferris Wheel* blackline master

Doing the Activity

Take a minute in class to go over the use of clock labels to represent Ferris wheel positions. (You may want to make a transparency of *The Ferris Wheel* blackline master.)

Discussing and Debriefing the Activity

In today's discussion, students do not need to develop any general formulas concerning the relationship between position and height. They will be working with that relationship further in *As the Ferris Wheel Turns* and several additional activities. For today, it is enough that students recognize that trigonometry will play a role. Keep in mind that from their current perspective, trigonometric functions exist only in a right-triangle context.

Let students discuss the problems for a few minutes in their groups, and have several groups prepare presentations for examples from Questions 2 and 3.

You can probably go over the various parts of Question 1 orally. These should be straightforward if students have read the information carefully. (They may have overlooked the fact that the low point of the Ferris wheel does not have height zero.)

Question 2

Have students do their presentations on Question 2. If they have trouble with these other positions on the Ferris wheel, you will need to review right-triangle trigonometry.

For example, for the 2 o'clock position, they should see that Al and Betty's height can be found using the triangle shown below. Thus, the height for the 2 o'clock position is given by the expression 5 + 15 + 15 sin 30°, which leads to a result of 27.5 feet.

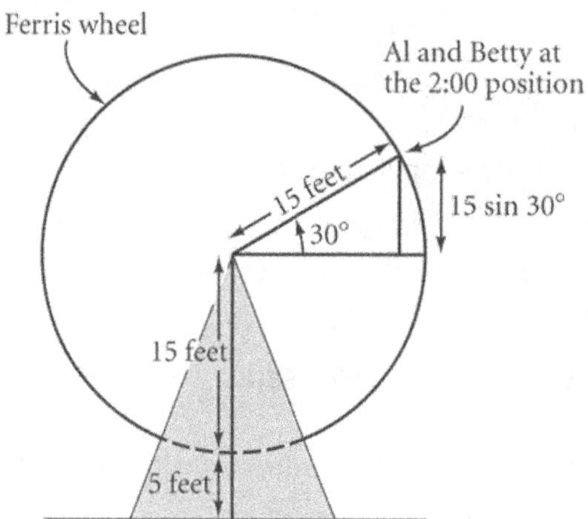

Students sometimes mistakenly assume that the heights are equally spaced from one hour to the next. For example, because the 3 o'clock position is 20 feet high and the 12 o'clock position is 35 feet high, students may think that the 2 o'clock position is a third of the way between them, at 25 feet. If this incorrect approach is presented, acknowledge that it is a reasonable idea and then have the class discuss its merits. Be sure students see that it gives the wrong result. The different "o'clock" positions create equal central angles, but they do not have equally spaced heights. (If no one brings up this incorrect approach, you might want to present it yourself, to get students to articulate why it doesn't work.)

Question 3

If students seemed clear about the ideas involved in their discussion of Question 2, you can omit discussion of Question 3. Otherwise, use one or two more examples to clarify the process of finding the height.

As the Ferris Wheel Turns

Intent

In this activity, students start to look at the relationship between time elapsed, clock position, and height on the Ferris wheel.

Mathematics

This activity requires students to find the speed of an object moving at a constant angular speed. They also find the height, for specific times, of an object moving in a circular path.

Progression

Students work on this activity individually and then discuss their results as a class.

Approximate Time

30 minutes for activity (at home or in class)

20 minutes for discussion

Classroom Organization

Individuals, followed by whole-class discussion

Doing the Activity

Students work on this activity independently.

Discussing and Debriefing the Activity

Note: The activity *At Certain Points in Time* depends on students really understanding this activity, so it's worth spending most of a class period on the discussion, even if that means that another activity is delayed.

You might begin by assigning a problem to each group to prepare for presentation. Then let different students report on their results.

Questions 1 through 3

For Question 1, students will need to use the circumference formula, $C = 2\pi r$, to see that the total distance traveled by the platform in one complete turn is 100π feet. Because the platform goes 100π feet in 40 seconds, it is going 2.5π feet per second, which is approximately 7.85 feet per second (roughly 5 miles per hour).

Post this result about the platform's speed, because it will be used later in the unit. You may want to incorporate it also into the posted diagram of the Ferris wheel:

> **The speed of the platform as it turns is 2.5π feet per second, which is approximately equal to 7.85 feet per second.**

On Question 2, students should see that because a complete turn is 360°, the Ferris wheel must be turning $360 \div 40 = 9$ degrees each second. Review the term *angular speed* (introduced in the activity), which measures how fast an angle is changing. Bring out that angular speed is measured in units such as degrees per second.

On Question 3, you may want to have a student identify each angle involved. Emphasize again that angular speed does not depend on the radius of the Ferris wheel.

Ask students, **What word is used to describe the time interval for each complete turn of the Ferris wheel?** If necessary, remind them of the term *period.*

Question 4

Question 4 is a lead-in to *At Certain Points in Time,* so take extra care to ensure that students understand the presentations. You may want to post the results from Question 4, because students will use this information in later activities.

Questions 4a and 4b are easier because they involve angles in the first quadrant. Use the presentations of these problems to help students work out the details of getting from the time elapsed to the angle of turn, as well as the process of using trigonometry to get the platform's height from the angle of turn. As students do various examples, they should see the need to distinguish the cases, depending on the quadrant.

For example, for $t = 6$, the situation looks like the diagram shown here. The angle is $6 \cdot 9°$ because the Ferris wheel turns 9 degrees per second. Students will probably use the equation $\sin 54° = \frac{y}{50}$ to get $y = 50 \sin 54°$. Therefore, the total height off the ground is $h = 65 + 50 \sin 54°$.

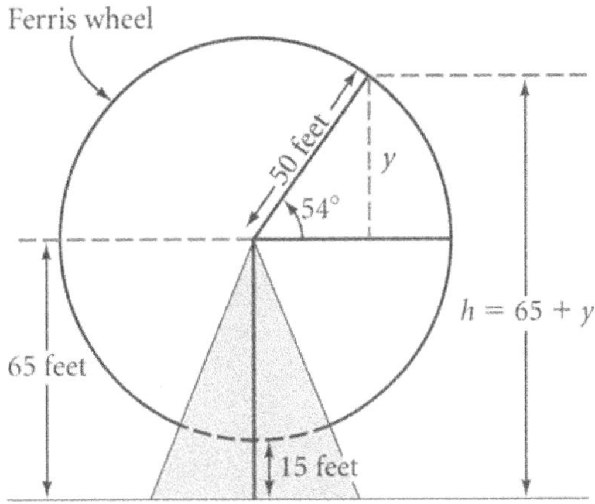

Students may recognize that Question 4c is the same as Question 1b of *The Ferris Wheel* except for the conversion from time to degrees.

For Question 4d (t = 14), the platform will have moved through an angle of 126°. Students may recognize that this leads to the same height for the platform as for Question 4b (t = 6). If they do not see this, they will probably use either the supplementary angle of 54° (getting 65 + 50 sin 54° for the height of the platform) or the related 36° angle (getting 65 + 50 cos 36°). These two approaches are illustrated in the following two diagrams. You don't need to discuss both methods, but, as always, encourage alternate approaches to problems. (*Note:* If both approaches come up, you can bring out that sin 54° = cos 36° and discuss why.)

Alternately, students may simply follow the pattern of Questions 4a and 4b, and use the expression 65 + 50 sin 126°. They may discover that this expression yields a reasonable answer on their calculators, even though sin 126° has not yet been defined. If this comes up, remind them that so far, they only know the meaning of the trigonometric functions for acute angles. Then tell them that extending the definitions of these functions is an important element in this unit.

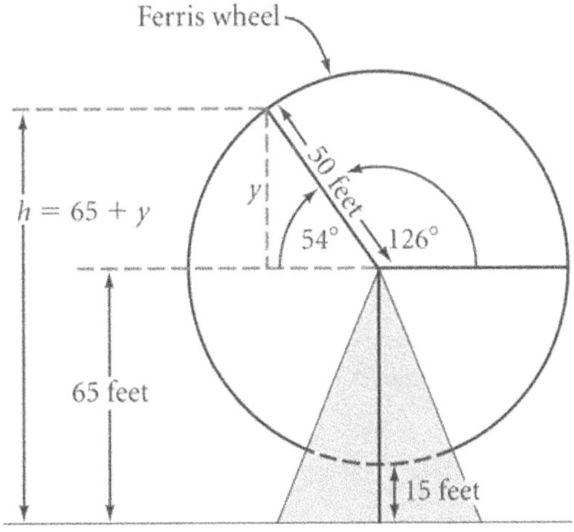

Here, we have $\frac{y}{50} = \sin 54°$, *so* $y = 50 \sin 54°$ *and* $h = 65 + 50 \sin 54°$.

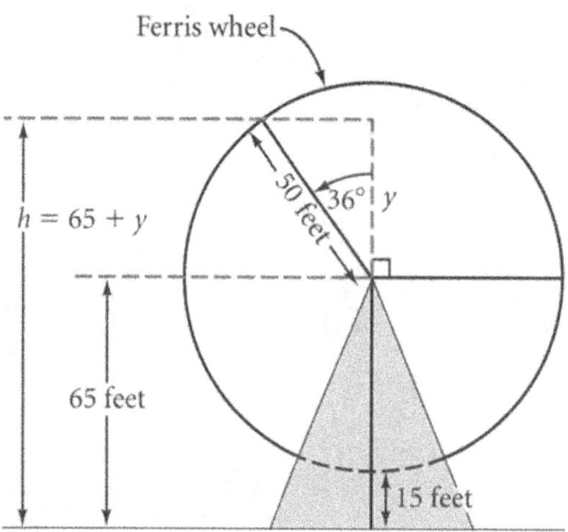

Here, we have $\frac{y}{50} = \cos 36°$, *so* $y = 50 \cos 36°$ *and* $h = 65 + 50 \cos 36°$.

For Question 4f, students will probably realize that when the Ferris wheel has turned for 49 seconds, the platform is at the same position as after 9 seconds. Use the word *periodic* to describe the motion of the platform.

Key Question

What word is used to describe the time interval for each complete turn of the Ferris wheel?

The Height and the Sine

Intent

In these activities, students explore the nature of the sine function.

Mathematics

The right-triangle definitions that were used to introduce students to the trigonometric functions are only relevant for acute angles. In this section, students extend the definition of the sine function to one that is meaningful for arbitrary angles. They graph sinusoidal functions, explore the periodic nature of the sine function, and use that periodicity to make sense of situations when their calculator returns only a principal value for the inverse sine.

Progression

Students begin by finding a general formula for the height of the central unit problem's Ferris wheel platform in the first quadrant (*At Certain Points in Time*). They obtain further familiarity with the relationship between the height of the platform and its angular position in *A Clear View*, then apply that knowledge to extend the definition of the sine function to angles beyond the first quadrant (*Extending the Sine*), including negative angles (*Testing the Definition*).

Students then explore the periodic nature of sinusoidal functions, first through the context of the Ferris wheel in *Graphing the Ferris Wheel* and *Ferris Wheel Graph Variations*, then looking at the graph of the simpler sine function in *The "Plain" Sine Graph*.

Sand Castles and *More Beach Adventures* challenge students to apply their new knowledge of sinusoidal functions to a situation that does not involve angles—the periodic rise and fall of the ocean tide. This situation becomes particularly thought provoking as students discover that even though the tides do not involve angles, they must resort to consideration of the principal angle of the inverse sine in order to make sense of the solutions to their equations.

At Certain Points in Time

A Clear View

Extending the Sine

Testing the Definition

Graphing the Ferris Wheel

Ferris Wheel Graph Variations

The "Plain" Sine Graph

Sand Castles

POW 15: Paving Patterns

More Beach Adventures

At Certain Points in Time

Intent

In this activity, students develop a formula for the platform's height when it is in the first quadrant.

Mathematics

In *As the Ferris Wheel Turns*, students found the height of the platform on the moving Ferris wheel at specific times. Now they generalize this process to find a formula for all heights in the first quadrant. At this point they are not able to extend their formula beyond the first quadrant, because they are only familiar with definitions of the trigonometric functions that are based on angles of right triangles, and thus only applicable to acute angles.

Progression

In this activity, students write a formula to generalize their findings from *As the Ferris Wheel Turns.* They'll extend their thinking beyond the first quadrant in *Extending the Sine.*

Approximate Time

25 minutes for activity

5 minutes for discussion

Classroom Organization

Small groups, followed by whole-class discussion

Doing the Activity

At Certain Points in Time asks students to represent the platform's height as h and to find a formula for h within the first quadrant in terms of the time in seconds (t) that the Ferris wheel has been turning. They are then asked to verify their formula using the results from Questions 4a and 4b of *As the Ferris Wheel Turns*.

Discussing and Debriefing the Activity

Let a student present and explain his or her group's formula. This should follow fairly easily from the discussion of *As the Ferris Wheel Turns*, but be sure to get a clear explanation, because this formula will play a major role in extending the sine function beyond acute angles.

Post and label the formula, which will probably look like this:

$$h = 65 + 50 \sin (9t)$$

Other quadrants will be considered in *Extending the Sine*.

A Clear View

Intent

In this activity, students continue to work with the relationship between the angular position of the platform and its height.

Mathematics

This activity uses Al and Betty's Ferris wheel and poses a question that isn't explicitly about only height or time. Students work backwards from the height of the platform to its angular position in order to determine the percentage of the time that the platform is above a fence.

Progression

Students work on this activity individually and then discuss their results as a class.

Approximate Time

30 minutes for activity (at home)

10 minutes for discussion

Classroom Organization

Individuals, followed by whole-class discussion

Doing the Activity

A Clear View asks students to find the percentage of the time that Al and Betty's Ferris wheel from *The Ferris Wheel* is above a 13-foot fence. They are then asked to explain how the answer would change if the period were different.

Discussing and Debriefing the Activity

Ask for a volunteer to present the problem. Be sure that the presenter includes a diagram as part of the explanation. For instance, the diagram here shows the situation when Al and Betty are exactly 13 feet off the ground (and in the fourth quadrant). At this position, they are 7 feet below the center of the Ferris wheel, so the angle θ satisfies the equation $\cos \theta = \frac{7}{15}$, which gives $\theta \approx 62.2°$. (You can use this discussion, if needed, as an opportunity to remind students of the inverse trigonometric functions and notation such as $\cos^{-1} \frac{7}{15}$.)

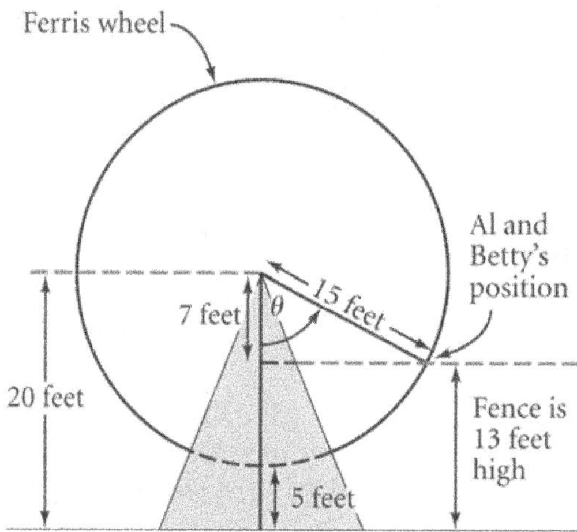

The presenter might use a diagram like the next one to bring out that Al and Betty are below the fence while the Ferris wheel travels through an angle of *twice* 62.2°, using the fact that there are two places in the Ferris wheel cycle where Al and Betty are 13 feet high.

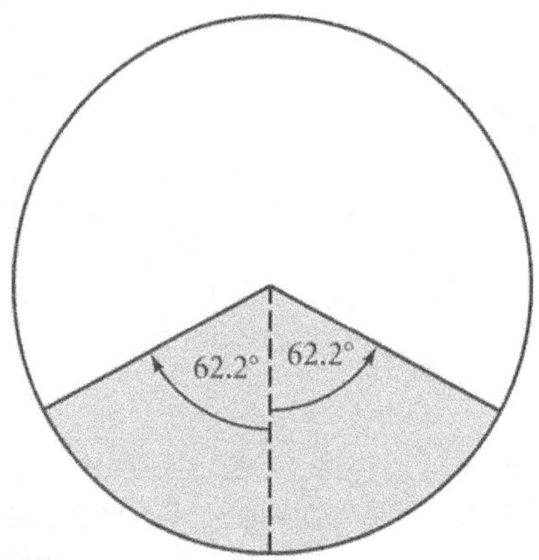

Some students may work directly with the angles. For instance, they might see that the fraction of the time during which Al and Betty are *below* the fence is equal to the ratio $\frac{2 \cdot 62.2}{360}$, or approximately 35%. Because Al and Betty are below 13 feet about 35% of the time, they can see over the fence about 65% of the time.

Some students may feel more comfortable working with the time involved. For instance, they might see that each second elapsed represents 15 degrees of turn, so 62.2 degrees represents $\frac{62.2}{15}$ seconds, or about 4.15 seconds. This means that Al and Betty are below the fence for about 8.3 seconds and above it for about

15.7 seconds. Thus, the fraction of the time that they are above the fence is about $\frac{15.7}{24}$, or approximately 65%.

You can use Question 2 to bring out that although the period of the Ferris wheel affects the *amount* of time in each cycle that Al and Betty are above the fence, it does not affect the *percentage* of time they are above the fence.

Extending the Sine

Intent

In this activity, students see how to extend the sine function to be defined for all angles.

Mathematics

Thus far, students have worked with trigonometric functions that have been defined in terms of right-triangle geometry and are thus applicable only to acute angles. Now they will see how to redefine the sine function in a manner that will make sense for all angles.

Progression

To begin this lesson, you'll remind students that the sine function has previously been defined only for first-quadrant angles, and students reflect on how the definition of exponentiation was previously extended beyond repeated multiplication in order to make sense of negative and zero exponents. Then, using a specific Ferris wheel situation, you'll illustrate the extended definition of sine, and students verify that calculator results agree with the new definition for specific cases.

Approximate Time

20 to 25 minutes

Classroom Organization

Teacher presentation

Doing the Activity

Extending the Sine summarizes the ideas behind extending the sine function beyond the first quadrant. Lead students through the discussion described below. Then, you might have students look over the *Extending the Sine* reference pages briefly, or you might have them begin work directly on the next activity, *Testing the Definition.*

Discussing and Debriefing the Activity

Review the work in *At Certain Points in Time* developing a first-quadrant formula for the platform's height above the ground,

$$h = 65 + 50 \sin (9t)$$

Point out that it would be nice to have a formula that works no matter where the platform is, and ask, **Why can't you simply apply this formula for all values of t?** Bring out that the definition of the sine function is based on right triangles, so the expression sin ($9t$) is meaningful only if $9t$ is strictly between 0° and 90°. That is, the formula makes sense (so far) only if the platform is still in the first quadrant, which means $0 < t < 10$. (Even if students have already discovered that the formula seems to work for other angles, point out that they don't have a general definition yet for the sine function.)

Previous Experiences with Extending Functions

Tell students that their next task is to consider how to extend the sine function to be defined for all angles. Ask, **When have you extended a function or operation before?** Remind them, if necessary, that when they first learned about exponentiation, the operation was defined in terms of repeated multiplication and that the definition made sense only if the exponent was a positive integer.

Then ask, **How did you extend exponentiation in** (the Year 2 unit) ***All About Alice*?** Try to bring out these key ideas:

- The new definition extending the domain of the operation had to be consistent with the old definition.
- The situation of Alice and the cake and beverage was a useful model for thinking about exponents.
- The new definition was created so that certain patterns and algebraic rules that held true for positive integer exponents continued to hold true when the domain was extended.

Tell students that the model of the Ferris wheel for circular motion will play a role for trigonometric functions similar to that played by the Alice situation for exponentiation

A Specific Case

Tell students that a key criterion of a new, extended definition of the sine function will be whether it makes the platform height function work for all values of t.

Point out that when $t = 14$, the height formula gives the expression $65 + 50 \sin (9 \cdot 14)$, which means that the angle of turn for the platform is 126°, so the platform would no longer be in the first quadrant. Ask, **What value should you assign as the definition of sin 126° so that the height formula gives the right answer for $t = 14$?** As a hint, you can ask, **What value of t gives the same height but leads to a first-quadrant angle?** Bring out that at $t = 14$, the platform has the same height as at $t = 6$, so the expression $65 + 50 \sin (9 \cdot 14)$

must evaluate to be the same as the expression 65 + 50 sin (9 · 6). To make this more clear, you might write this as the equation

$$65 + 50 \sin 126° = 65 + 50 \sin 54°$$

Thus, for the height formula to work for $t = 14$, we must define the sine function so that sin 126° has the same value as sin 54°.

Defining the Sine Function for All Angles

Tell students that in order to create a general, extended definition of the sine function, it's helpful to replace the Ferris wheel model with the more abstract setting of the coordinate plane. You can begin with a diagram like the one shown here, suggesting that they imagine that the circle of the Ferris wheel has been placed in the coordinate plane, with its center at the origin.

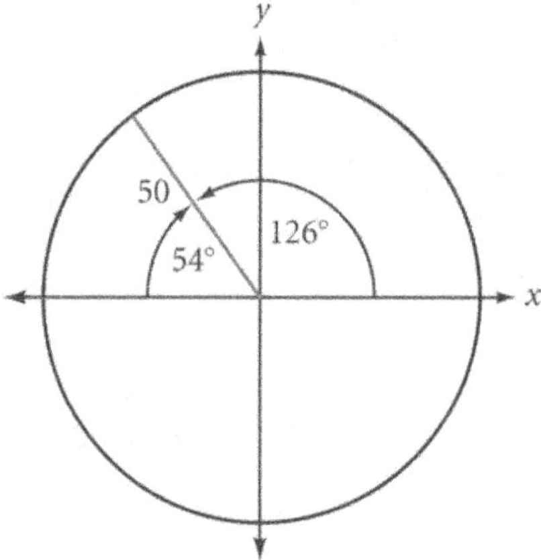

Then ask the class to consider the next diagram, which corresponds to a situation in which the Ferris wheel has turned only through a first-quadrant angle. A generic point *A*, with coordinates (*x*, *y*), has been marked on the ray defining the angle. (The point *A* is assumed to be different from the origin.)

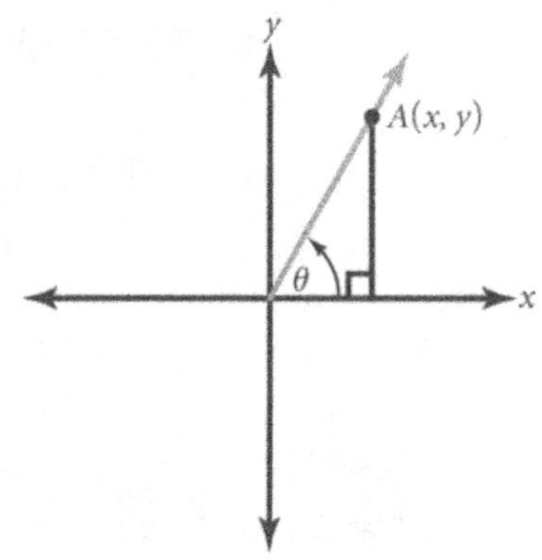

Ask, **How can sin θ be defined in terms of the coordinates x and y?** Students should see that in the right triangle, sin θ is equal to the ratio $\frac{y}{\sqrt{x^2 + y^2}}$.

Introduce the use of r as a shorthand for the expression $\sqrt{x^2 + y^2}$, and bring out that this corresponds to the radius of the Ferris wheel. Use the variable r to rewrite the expression for sin θ more simply as the ratio $\frac{y}{r}$.

Point out that the ratio $\frac{y}{r}$ makes sense for any angle. (The issue of negative angles will be discussed explicitly following *Testing the Definition*.) Tell students that this simple ratio is used for the extended definition of the sine function. Post the formal definition together with an appropriate diagram:

> **For any angle θ, we define sin θ by first drawing the ray that makes a counterclockwise angle θ with the positive x-axis and choosing a point A on this ray (other than the origin) with coordinates (x, y).**

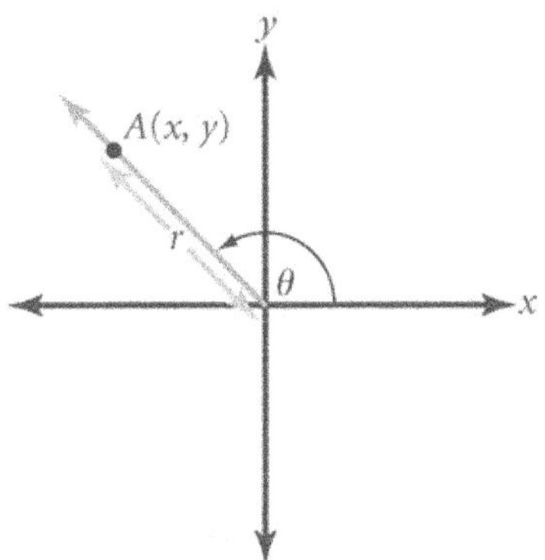

> **Using the shorthand $r = \sqrt{x^2 + y^2}$, we then define the sine function by the equation**
>
> $$\sin \theta = \frac{y}{r}$$

Note: The issue of why this method makes sin θ well-defined is discussed following *Testing the Definition* (see the section "Why Is the Sine Function Well Defined?").

This extension of the right-triangle sine function to the more general definition of sine for arbitrary angles is a key idea, and perhaps even the central mathematical idea of this unit, so be sure to give it appropriate emphasis.

Bring out that we have taken familiar ideas—right-triangle trigonometry and the coordinate system—and combined them in the context of a concrete situation to create a more general definition of the sine function. This new definition is consistent with the old definition of sine for acute angles. It allows us to replace a complex, quadrant-by-quadrant analysis of the platform height with a single, uniform expression.

Back to the Specific Case

Have students work in groups and ask, **How does this definition apply to the case $\theta = 126°$?** You may need to suggest that they pick a specific value for r (perhaps $r = 50$, as in the Ferris wheel) and then find the corresponding value for y, as in this diagram:

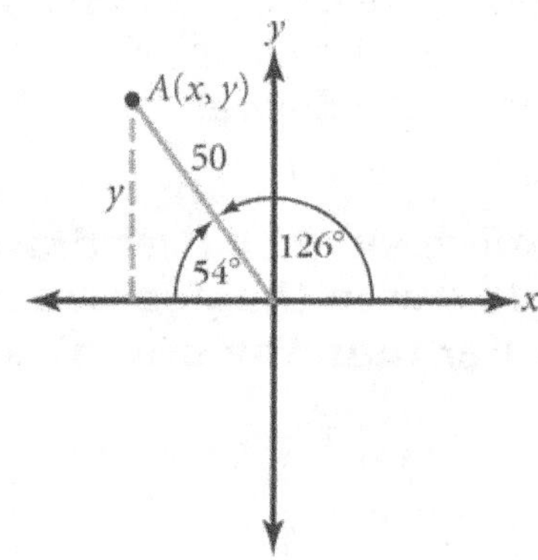

Students should see, perhaps using the right triangle in the second quadrant, that y is equal to 50 sin 54°, so the ratio $\frac{y}{r}$ is equal to sin 54°. In other words, this coordinate definition will make sin 126° the same as sin 54°, as before.

Have students verify that their calculators do, in fact, give the same value for sin 126° as for sin 54°.

Key Questions

Why can't you simply apply this formula for all values of *t*?

When have you extended a function or operation before?

How did you extend exponentiation in *All About Alice*?

What value should you assign as the definition of sin 126° so that the height formula gives the right answer for $t = 14$?

What value of *t* gives the same height but leads to a first-quadrant angle?

How can sin θ be defined in terms of the coordinates x and y?

How does this definition apply to the case $\theta = 126°$?

Testing the Definition

Intent

In this activity, students verify that the extended definition of the sine works in the platform height formula for all angles.

Mathematics

In *Extending the Sine*, students were presented with the development of a new definition for the sine function that makes sense for arbitrary angles rather than only for acute angles. Now they will test that definition in the formula for the height of the platform developed in *At Certain Points in Time* to see if it will successfully extend that formula to all four quadrants. The discussion will also introduce reference angles and the unit circle and will look at the case of the sine of a negative angle.

Progression

Students work on the activity in groups and then discuss their results as a class. The discussion also introduces new concepts—the unit circle, reference angles, and sine of a negative angle.

Approximate Time

20 minutes for activity

35 minutes for discussion

Classroom Organization

Small groups, followed by whole-class discussion

Doing the Activity

Question 1 uses the special case in which the right triangle has a 45° angle. On Question 2, students will need to use right-triangle trigonometry to find the coordinates of some point A in the fourth quadrant.

You need not have all groups complete Question 2 before beginning discussion.

Discussing and Debriefing the Activity

Let students from different groups present different parts of Question 1. For Question 1a, students might pick 50 for r and see that the legs of the right triangle are both of length $25\sqrt{2}$. They might get this by using the Pythagorean theorem, by using trigonometric functions of a 45° angle, or simply by remembering the ratios for the special case of an isosceles right triangle.

Once students have the lengths of the sides, they need to see that y is negative, because point A is in the third quadrant. (*Comment:* They don't actually need to find the value of x to get the value of sin 225°, but you might have them do this anyway.)

Finally, students need to find the ratio $\frac{y}{r}$ and see that this ratio is approximately −0.707. (If they leave the ratio as $-\sqrt{2}$, that's fine, too.)

Questions 1b through 1d

When students substitute their value for sin 225° into the expression 65 + 50 sin 225°, they should get a value of approximately 29.6 feet. Their explanation that this is reasonable should certainly include the fact that the result is less than 65 feet (because the platform is below the center of the Ferris wheel) but more than 15 feet (because the lowest point in the cycle is 15 feet off the ground).

You might ask them to find the position of the platform after 5 seconds (which corresponds to an angle of 45°). Then bring out that the result there (about 100.4 feet) is about 35.4 feet above the center of the Ferris wheel, just as 29.6 feet is 35.4 feet below the center.

Be sure to have students see that their calculators give the same value for sin 225° as the value they just found by hand.

Question 2

Question 2 is similar to Question 1, except that students will likely have more difficulty finding the value of y here than in Question 1. Ask, **How can you express sin 288° in terms of the sine of a first-quadrant angle?** They should see that except for sign, sin 288° is like sin 72°, so sin 288° = −sin 72°. As with Question 1, have them see that their calculators give this result also.

The Reference Angle

Bring out that for any angle, there is a first-quadrant angle that has the same sine, except perhaps for the sign. (Be careful to distinguish between the words "sign" and "sine.")

Tell students that this first-quadrant angle is called the *reference angle.* They should see that the reference angle for 225° (from Question 1) is 45° and that the reference angle for 288° is 72°.

The General Platform Height Function

Assure students that with the coordinate definition of the sine function, their platform height function works for all values of t. Whenever the platform is above the center of the Ferris wheel, y is positive; whenever the platform is below the center of the Ferris wheel, y is negative. Thus, the platform's overall height is $65 + y$, regardless of the sign of y.

By the general definition, $\sin \theta = \frac{y}{r}$, so $y = r \sin \theta$. In the case of the Ferris wheel in the unit problem, y becomes $50 \sin \theta$. In other words, the height of the Ferris wheel is always $65 + 50 \sin \theta$. That is the power of using the coordinate plane to understand the Ferris wheel.

Post this general result, which you might state as follows:

Suppose the Ferris wheel and platform satisfy these conditions:

- **The center of the Ferris wheel is 65 feet off the ground.**
- **The radius of the Ferris wheel is 50 feet.**
- **The Ferris wheel turns counterclockwise at a constant rate with a period of 40 seconds.**
- **The platform is at the 3 o'clock position when $t = 0$.**

Then the height of the platform off the ground after t seconds is given by the expression

$$65 + 50 \sin (9t)$$

Be sure students see how the specific details of the circus act fit into this formula. In particular, you may want to go over the fact that the coefficient 9 (in the expression $9t$) comes from dividing 360° by the period of 40 seconds. You might review the term *angular speed* by bringing out that the angular speed of the Ferris wheel is 9 degrees per second.

Negative Angles

Ask, **How might the extended definition of the sine function make sense when θ is negative?** You may want to suggest that students consider what a negative value for t would mean in the context of the Ferris wheel problem.

Students may be able to guess on their own, but if needed, tell them that we interpret a negative angle by going clockwise instead of counterclockwise and that a negative value of *t* refers to a time before the platform reaches the 3 o'clock position.

Illustrate with a specific case, such as asking students to find sin (–53°). They should see that the reference angle is 53° and that sin (–53°) = –sin 53°.

Why is the Sine Function Well Defined?

The definition of sin θ involves the use of a point on the ray making an angle of θ measured counterclockwise from the *x*-axis. The question may have come up earlier as to why we can use *any* point on the ray. If it has not yet been discussed, bring it up now.

Review the fact that sin θ is defined by the ratio $\frac{y}{r}$ for some point (other than the origin) on the ray defining θ. You might use a diagram like this:

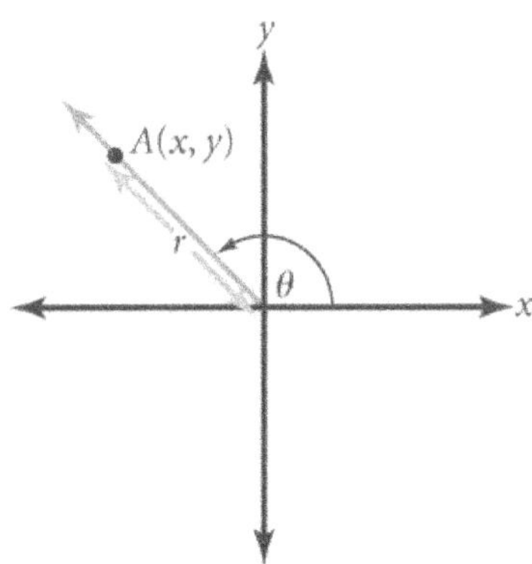

Then ask, **Why is the ratio $\frac{y}{r}$ the same for all points on the ray?** If necessary, revise the diagram to show two points on the ray, perhaps labeled (x_1, y_1) and (x_2, y_2), with *r*-values r_1 and r_2, like this:

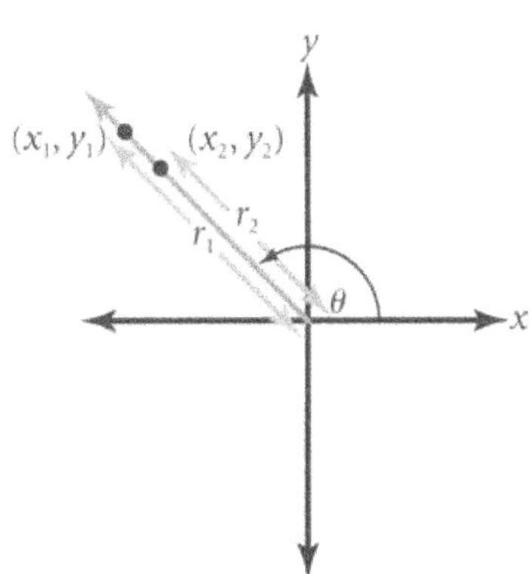

If needed, restate the question in terms of the new diagram, asking, **Why is the ratio $\frac{y_1}{r_1}$ the same number as $\frac{y_2}{r_2}$?** The goal is to bring out that similarity plays a key role in this general definition, just as it did when the trigonometric functions were defined for acute angles using right triangles. You might point out that no matter what quadrant θ is in, the two ratios will have the same sign because the *y*-coordinates of the two points have the same sign. Students can then use similar right triangles to see that the ratios of the lengths of the sides are the same by similarity.

The Unit Circle

Tell students that because the value of the sine does not change for different values of *r*, they can choose whatever value is most convenient. Ask, **What value for *r* would be simplest to use?** Bring out that using 1 for *r* means simply that $\sin \theta = y$.

Ask, **How can you describe the set of points with $r = 1$?** Students should see that these points form a circle of radius 1, with the center at the origin. Tell them that this set of points is called the **unit circle.**

Key Questions

How can you express sin 288° in terms of the sine of a first-quadrant angle?

How might the extended definition of the sine function make sense when θ is negative?

Why is the ratio $\frac{y}{r}$ the same for all points on the ray?

Why is the ratio $\frac{y_1}{r_1}$ the same number as $\frac{y_2}{r_2}$?

What value for *r* would be simplest to use?

How can you describe the set of points with $r = 1$?

Graphing the Ferris Wheel

Intent

In this activity, students graph the Ferris wheel height function.

Mathematics

Having extended the definition of the sine beyond angles of the first quadrant, students now graph the Ferris wheel height function through two full revolutions, observing the periodicity of the graph. They then think about how changing the parameters affects the graph.

Progression

Students work on the activity individually and then discuss their results as a class.

Students will explore in more detail how changing parameters affects the graph in *Ferris Wheel Graph Variations.*

Approximate Time

30 to 40 minutes for activity (at home or in class)

15 minutes for discussion

Classroom Organization

Individuals, followed by whole-class discussion

Materials

Transparencies of the *Graphing the Ferris Wheel* blackline masters

Doing the Activity

You might suggest to students that they use their data from *As the Ferris Wheel Turns* in Question 1 of this activity.

Discussing and Debriefing the Activity

Regarding the graph in Question 1, ask, **How should the axes and scales be set up?** The vertical scale should reflect the fact that the height goes from a minimum of 15 feet to a maximum of 115 feet. Students are specifically instructed to include a horizontal scale from $t = 0$ to $t = 80$.

You might display a transparency with these scales so that students can plot their points on a shared coordinate system. (A blackline master of a blank coordinate system with these scales is included in the *Graphing the Ferris Wheel* blackline masters.) You can then have students from various groups each give and explain the coordinates for one point on the graph. You can label various points with the expressions used to find the values for h.

Some students may be finding the heights in terms of right triangles, just as they did at the beginning of the unit, and perhaps even using the cosine of some angle rather than the sine. If so, you may get a diagram with points labeled as shown here. (The *Graphing the Ferris Wheel* blackline masters include a larger version of this graph without individual points labeled.)

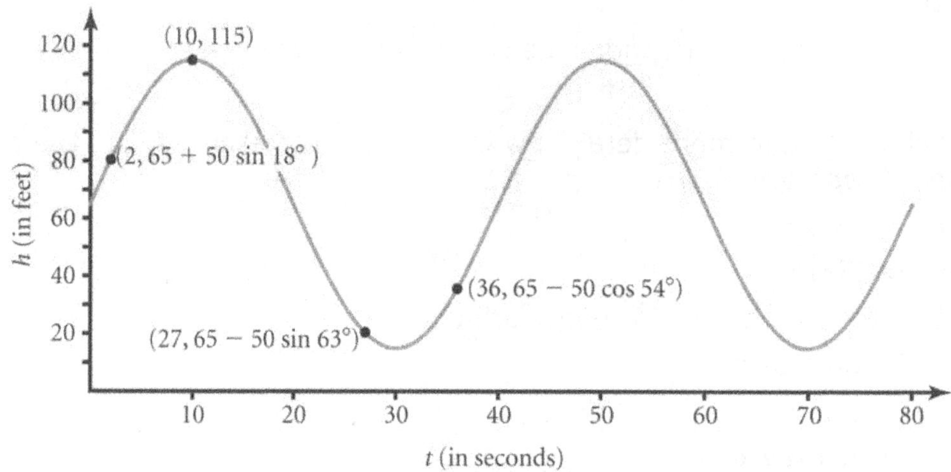

Ask, **What single equation will describe this graph?** If needed, review the discussion from *Testing the Definition*, so students see that this is the graph of $h = 65 + 50 \sin (9t)$.

Have students graph this function on their calculators (adjusting the variables as needed for calculator entry and choosing an appropriate viewing window). They should see that the graph matches the one they created by hand.

Symmetry and Periodicity in the Graph

Ask, **Did you use any shortcuts or see any patterns?** Guide them to mention things like symmetry or periodicity here.

This is a good time to review the idea of periodicity. Ask, **What does it mean that f is periodic, with period 40 seconds?** Students might say something like, "The height is the same every 40 seconds." You can have a student use the graph to illustrate what this means, by showing that points whose *t*-coordinates differ by 40 have the same *h*-coordinate.

Bring out that when we say that the period for the height function is 40 seconds, this means not only that the height is the same every 40 seconds but also that there is no smaller time interval for which the height always repeats.

Question 2

The discussion of Question 2 should be limited to a qualitative description of the changes in the graphs. In the next activity, *Ferris Wheel Graph Variations*, students will look at the graphs for specific variations of each type.

In discussing Question 2a, students should be able to explain that if the radius were smaller, the new graph would not go as "high" or as "low" as the original. They might describe the new graph as "squished vertically toward the line $y = 65$." Bring out that the "midline" of the graph remains the same. That is, the graph is still as much above the line $y = 65$ as it is below this line.

Tell students that the distance from the midline to the high or low point of the graph is called the **amplitude** of the graph. In other words, the amplitude for such a Ferris wheel height graph is the same as the radius of the Ferris wheel.

On Question 2b, students should see that if the Ferris wheel turns faster, then the platform will go up and down more times during the 80-second interval shown. In other words, the height function will have a smaller period. They might describe the graph as "squished horizontally like an accordion."

Finally, on Question 2c, students should see that the graph is simply moved down so that it is above the axis half the time and below the axis half the time. Ask, **How is the amplitude affected in Question 2c?** Students should see that the amplitude has not changed.

Key Questions

How should the axes and scales be set up?

What single equation will describe this graph?

Did you use any shortcuts or see any patterns?

What does it mean that *f* is periodic, with period 40 seconds?

How is the amplitude affected in Question 2c?

Ferris Wheel Graph Variations

Intent

In this activity, students examine how changing the specifications of the Ferris wheel affects the graph of the platform's height.

Mathematics

In *Graphing the Ferris Wheel*, students were asked to describe how various changes in the parameters of the Ferris wheel would affect the graph of the platform's height function. Now they validate their intuitive responses by creating new graphs for the height function with different specific values for the parameters. They also look at how the equation describing the function changes.

Progression

Students work on the activity individually and share their results as a class. This activity formalizes the intuitive observations that students made in *Graphing the Ferris Wheel.*

Approximate Time

30 minutes for activity (at home or in class)

15 minutes for discussion

Classroom Organization

Individuals, followed by whole-class discussion

Doing the Activity

The first two questions in *Ferris Wheel Graph Variations* ask students to pick a new value for a specified parameter of the Ferris wheel, draw a new graph of the height function, give an equation for the graph, and verify that the equation works for a specific value of t. The final question is similar, with the Ferris wheel in a hole such that its center is at ground level.

Discussing and Debriefing the Activity

You can have volunteers present specific graphs for each of the questions. Keep in mind that other students may have chosen different values for the radius or period, so students will probably have slightly different graphs. You can use the specific examples to review the general ideas that were included in the discussion of *Graphing the Ferris Wheel*.

Question 1

If the presenter of Question 1 does not use the term *amplitude*, ask, **How does the amplitude of this graph compare to that of the original graph?** Students should see that the amplitude of the graph for Question 1 is less than that of the "original" platform height graph (from *Graphing the Ferris Wheel*).

More generally, ask, **What does the amplitude (for a Ferris wheel height graph) depend on?** Bring out that the amplitude of the graph is equal to the radius of the Ferris wheel.

Connect the changes in the Ferris wheel and the graph to the change in the equation. For instance, if a student changed the radius to 30 feet, the equation would be $h = 65 + 30 \sin(9t)$.

Ask, **Has anything happened to the period?** Students should see that the period is not affected by the change in radius.

Question 2

For Question 2, ask, **Has anything happened to the amplitude?** Students should see that the amplitude is not affected by the change in period.

As with Question 1, connect the change in the Ferris wheel and the graph to the change in the equation. For instance, if a student changed the period to 20 seconds, the coefficient of t would change from 9 to 18 (found by dividing 360° by 20), and the equation would be $h = 65 + 50 \sin(18t)$.

Question 3

For Question 3, students should all have the same graph and equation. They should see that the entire graph is merely moved down so that the x-axis is its midline. The equation is simply $h = 50 \sin(9t)$.

Ask, **Has anything happened to the period or amplitude?** Students should see that they have not changed.

Key Questions

How does the amplitude of this graph compare to that of the original graph?

What does the amplitude depend on?

Has anything happened to the period?

Has anything happened to the amplitude?

Has anything happened to the period or amplitude?

Supplemental Activity

A Shifted Ferris Wheel (extension) guides students to investigate the effect of changing the initial location of the platform from the 3 o'clock position to something else.

The "Plain" Sine Graph

Intent

In this activity, students graph the sine function and observe characteristics of the graph.

Mathematics

In the last two activities, students worked with graphs of "Ferris wheel situations" involving the sine function. In this activity, students will examine the graph of the sine function itself.

Progression

Students work on this activity in groups, and discuss their results as a class. The discussion following this activity also points out the graph's amplitude, period, intercepts, and maxima and minima.

Approximate Time

25 minutes for activity

10 minutes for discussion

Classroom Organization

Small groups, followed by whole-class discussion

Materials

Transparency of *The "Plain" Sine Graph* blackline master

Doing the Activity

Students should be able to do this activity with no introduction.

Discussing and Debriefing the Activity

The discussion of this activity can be brief if students were comfortable answering the specific questions in their groups.

You should post a copy of the graph itself. (*The "Plain" Sine Graph* blackline master provides a graph similar to the diagram here, but with domain -180° to 540°.)

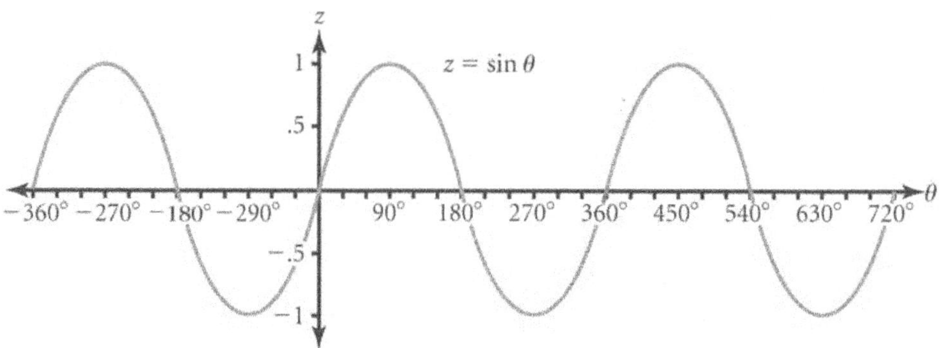

Some Further Details

Here are some specific connections and ideas to bring out, if they do not seem clear from students' group work:

- Have students explain in terms of the coordinate definition why the sine function has a maximum of 1 and a minimum of −1.
- Bring out the connection between the facts that the amplitude of the function is 1 and that the sine function has a maximum of 1 and a minimum of −1.
- Have students write an equation expressing that the period of the function is 360°. For instance, they might write $\sin(\theta + 360°) = \sin\theta$ and explain this in terms of the coordinate definition.
- Have students relate the θ-intercepts to the coordinate definition of the sine function. They should see that the sine function is zero when the "defining point" (that is, the point on the appropriate ray) has a y-coordinate of zero, which means that the point is on the x-axis.
- Bring out that the angles that are involved in the intercepts, maximum points, and minimum points are precisely the angles for which the defining point is on one of the axes, so that y is either zero or is equal to r in absolute value.

The Sign of the Sine

Ask, **For what angles is the sine function positive? When is it negative?** By looking at the sign of the y-coordinate, students should be able to determine that the sine function comes out positive if the angle is in the first or second quadrant and negative if the angle is in the third or fourth quadrant. (Bring out that an angle such as −53° is a fourth-quadrant angle.)

You may find it helpful to use a diagram like this one,

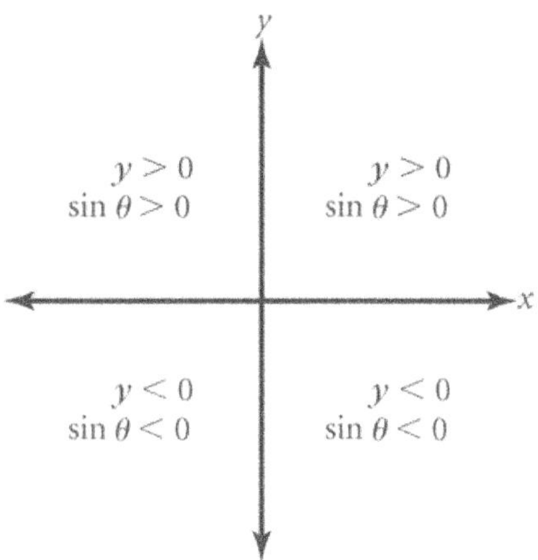

The Sine Function and the Ferris Wheel

Finally, bring out that this graph has the same basic shape as the "height functions" that students examined in *Ferris Wheel Graph Variations*. Ask, **What specifications for the Ferris wheel would give this graph?** They should see that this graph shows the height of the platform for a Ferris wheel with a radius of 1 unit and an angular speed of 1 degree per second and for which the center of the wheel is at ground level.

Key Questions

For what angles is the sine function positive? When is it negative?

What specifications for the Ferris wheel would give this graph?

Sand Castles

Intent

In this activity, students apply the extended sine function in a new problem situation.

Mathematics

This activity illustrates the use of a sinusoidal function to describe periodic motion in a context that is quite different from the Ferris wheel and that does not involve angles in any obvious way.

Progression

Sand Castles describes the tide action on a beach as fitting a given sine function and asks students to answer several questions related to the location of the waterline at various points in time. This activity is usually difficult for students and sparks substantial class discussion.

Approximate Time

40 to 50 minutes for activity (at home or in class)

25 to 30 minutes for discussion

Classroom Organization

Individuals, followed by whole-class discussion

Doing the Activity

Students work on this activity independently.

Discussing and Debriefing the Activity

You may want to give students a minute to check their work using a graphing calculator before you begin the discussion. They might use the "trace" feature or a table to check for maximum and minimum points and to confirm results for Questions 2 through 5.

Have volunteers present their results on each question.

Question 1

In the discussion of Question 1, have the presenter explain how he or she made the graph. For instance, the student may have realized that the first maximum occurs when $29t$ is equal to 90 (which gives $t \approx 3.1$) and that the water is at its average value again when $29t$ is equal to 180 (which gives $t \approx 6.2$). Help the class to recognize the connection between the water level graph and the graph of the "plain" sine function.

Bring out that a 24-hour period will cover slightly less than two full periods for the water's motion. (You might ask at this point what the period of the function is, although that could also come out in connection with Question 3.)

The graph might look like the diagram here, although students might also label their scales to indicate where the maximum and minimum points are, rather than show whole numbers of hours. (They also might use a 24-hour interval other than from $t = 0$ to $t = 24$.)

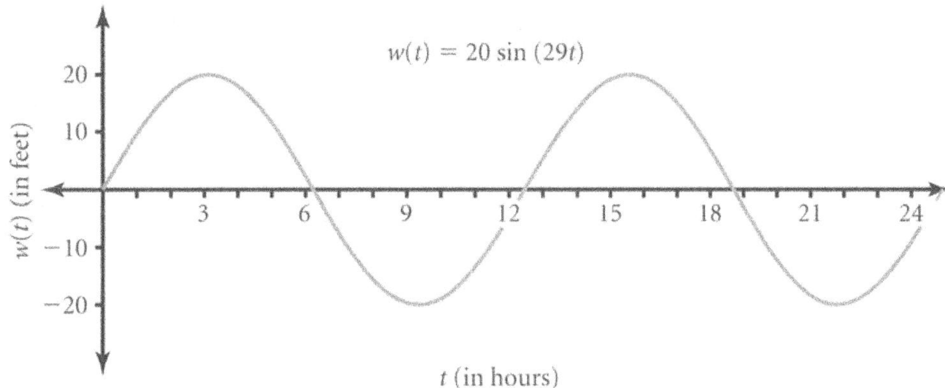

Question 2

For Question 2, students should see that the water level will get as high as 20 feet above and as low as 20 feet below the average waterline. Have them explain how they can see this at a glance from the function, and identify the number 20 as the amplitude of the function. (*Note:* The maximum value for $w(t)$ occurs at approximately $t = 3.1$ and 15.5, and the minimum at approximately $t = 9.3$ and 21.7. In other words, high tide is at about 3:06 a.m. and 3:30 p.m., and low tide is at about 9:18 a.m. and 9:42 p.m.)

Question 3

On Question 3, have the presenter give the specific times when the waterline is at its average level and identify these times on the graph of the function. Students should see that the question refers to the duration of the "lower half" of the curve. Be sure they identify the length of this time interval (which is about 6 hours and 12 minutes) as half the period of the function.

Question 4

For Question 4, the presenter will probably have looked for solutions to the equation $w(t) = -10$, which simplifies to $\sin(29t) = -0.5$. This is a good opportunity to review the idea of the inverse sine function and to discuss its limitations. (The more general inverse sine function will be discussed in *More Beach Adventures.*)

Students should see that their calculators give –30 as the value for $\sin^{-1}(-0.5)$. Discuss what this means, bringing out that the inverse sine function, as a function, can give only one of the angles whose sine is –0.5.

Ask, **How can you use the fact that $\sin(-30°) = -0.5$?** Students might use a graph, the Ferris wheel, or a coordinate system diagram to find other angles whose sine is –0.5.

In this problem, students are looking for two times where the waterline is at –10 feet: one (as the tide goes out) on the "down side" of the graph and one (as the tide comes back in) on the subsequent "up side" of the graph, as shown in the following diagram. They should see that these two times could correspond to solutions of the equation $20 \sin(29t) = -10$ with $29t$ representing angles in the third and fourth quadrants. This leads to the conditions $29t = 210°$ and $29t = 330°$, because both sin 210° and sin 330° are equal to –0.5. (As the diagram indicates, Oceana could also use the next tide cycle, which gives the conditions $29t = 570°$ and $29t = 690°$.)

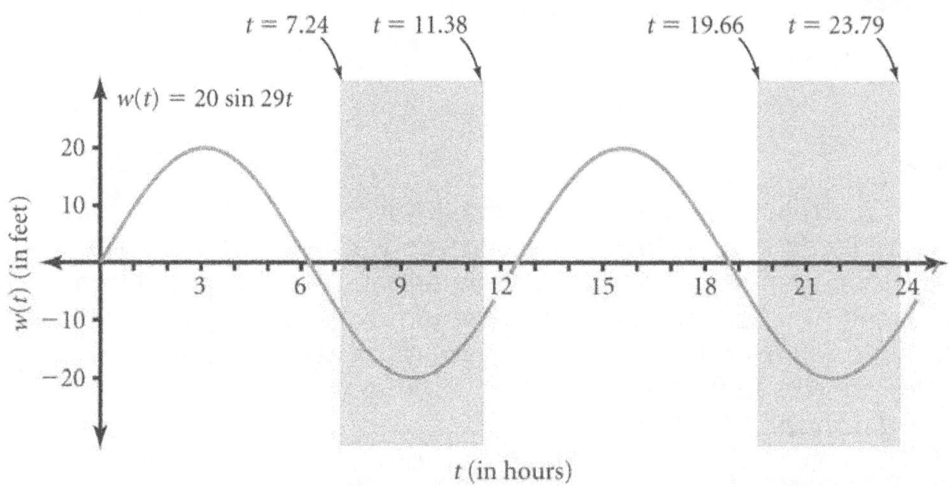

Shaded areas represent Oceana's options for the time interval in Question 4.

In other words, the water level reaches 10 feet below average about 7.24 hours after midnight (or at about 7:14 a.m.) and reaches that level again about 11.38 hours after midnight (or at about 11:23 a.m.). This gives Oceana about 4.14 hours to construct her sand castle. (The equations $29t = 570°$ and $29t = 690°$ give Oceana from about 7:39 p.m. to about 11:48 p.m., which is again a time interval of about 4.14 hours.)

Question 5

Question 5 is somewhat similar to Question 4. Again, students might use a graph or other diagram to get a sense of what's going on, perhaps seeing that Oceana needs to determine the water level one hour before low tide. They might use their knowledge about the period to see that the low tide first occurs about 9.31 hours after midnight, so Oceana should start about 8.31 hours after midnight. They then need to find 20 sin (29 · 8.31), which is approximately −17.5. In other words, if Oceana wants two hours in which to build her castle, the lowest she can go is 17.5 feet below the average waterline.

Sines without Angles

If it hasn't yet come up, you may want to point out that the central equation in this problem involves the sine function, but the problem as stated has nothing to do with angles. Tell students that the type of "rise and fall" motion that is involved in this problem and that is shown in the graph of the sine function occurs in many contexts that do not involve angles.

Key Question

How can you use the fact that sin (−30°) = −0.5?

POW 15: Paving Patterns

Intent

In this activity, students solve a complex problem requiring substantial exploration and write a clear justification of their solution.

Mathematics

This POW involves the Fibonacci sequence and will give students another opportunity to work with recursion.

Progression

After a brief introduction, give students about a week to work on the POW.

Approximate Time

5 minutes for introduction

1 to 3 hours for activity (at home)

25 to 35 minutes for presentations and discussion

Classroom Organization

Individuals, followed by several student presentations and whole-class discussion

Doing the Activity

POW 15: Paving Patterns asks students to find the number of ways that 1-foot by 2-foot tiles could be arranged in a 2-foot by 20-foot path, and suggests that they begin by looking at paths with shorter lengths. They are also asked to give any general formulas they found and any explanations for those formulas.

As students compile their data for paths of varying lengths, the tabulated results form a sequence similar to the Fibonacci sequence, which is most easily generalized with a recursion equation.

Allow students about a week to work on this POW. On the day before it is due, choose three students to make POW presentations on the following day, and give them overhead transparencies and pens to take home to use for preparing those presentations.

Discussing and Debriefing the Activity

Let the three students make their presentations, and have other students add their ideas.

If students make a table of results, showing the number of ways to pave an $n \times 2$ path, it will look something like this:

Length of path	Number of paving patterns
1 foot	1
2 feet	2
3 feet	3
4 feet	5
5 feet	8
6 feet	13

For instance, here are the eight different ways in which a 5×2 rectangle can be paved:

There are several important aspects to this problem. The first stage is students' ability to organize their lists of paving arrangements so that they get the right data in the table. It may be productive for them to share ways in which they avoided omitting any patterns.

The second stage is recognizing the pattern in the table. The sequence 1, 2, 3, 5, 8, 13, . . . is a slight variation on the sequence known as the *Fibonacci sequence.* (The Fibonacci sequence begins with two 1's instead of one. That is, it goes 1, 1, 2, 3, 5, 8, and so on.)

Historical Note: About Leonardo Fibonacci

Leonardo Fibonacci (c. 1170–1240), also known as Leonardo of Pisa, was a major mathematician of the Middle Ages. He studied with an Arab master while his father served as consul in North Africa. In his first book, *Liber Abaci* (Book of the Abacus), published in 1202, he made the Hindu-Arabic numeral system—the base 10 place value system—generally available in Europe. Prior to that time, the system was known in Europe only to a few intellectuals who had seen translations of the writings of the ninth-century Arab mathematician and astronomer al-Khwarizmi.

Liber Abaci also contained a discussion of the number sequence that now bears Fibonacci's name, introducing the sequence in connection with this problem: A certain man put a pair of rabbits in a place surrounded on all sides by a wall. How many pairs of rabbits can be produced from that pair in a year if it is supposed that every month each pair begets a new pair which from the second month on becomes productive?

The Pattern in the Table

As students will probably see, each number in the right-hand column of the table is the sum of the two preceding numbers. For instance, the entry 13 is the sum of the two preceding terms, 8 and 5.

Formally, if we let a_n represent the number of ways to pave an $n \times 2$ rectangle, then the pattern in the table can be represented by the formula

$$a_n = a_{n-1} + a_{n-2}$$

For example, the case $n = 6$ gives us the formula $a_6 = a_5 + a_4$, which is our earlier relationship $13 = 8 + 5$. (Because a_n is defined only for positive values of *n,* the formula $a_n = a_{n-1} + a_{n-2}$ makes sense only if $n \geq 3$.)

Ask, **What is this type of formula called?** Students may recall this from *POW 14: The Tower of Hanoi,* but if necessary, remind them of the term *recursive formula.*

Ask, **What does this formula say about the paving patterns?** Bring out that it says that the number of *n* x 2 paths is the sum of the number of (*n* - 1) x 2 paths and the number of (*n* - 2) x 2 paths.

Students can continue the table using the recursive formula. For instance,

$$a_7 = a_6 + a_5 = 13 + 8 = 21$$
$$a_8 = a_7 + a_6 = 21 + 13 = 34$$
$$a_9 = a_8 + a_7 = 34 + 21 = 55$$

They can continue in this way to get the equation

$$a_{20} = a_{19} + a_{18} = 6765 + 4181 = 10{,}946$$

Thus, there are nearly 11,000 ways for Al and Betty to pave a 20 × 2 path with their 1 × 2 paving stones.

Explaining the Recursive Formula

An important aspect of the problem is understanding, **Why does this recursive formula hold true?** Ask the class why the number of n x 2 paths should be the sum of the number of (n – 1) x 2 paths and the number of (n – 2) x 2 paths. If no one can explain this, here is a sequence of questions you can ask to lead students to understand the pattern.

First, ask, **What can the 'beginning' of the paving pattern look like?**, bringing out that there are two options. One option is to place a single stone sideways, like this:

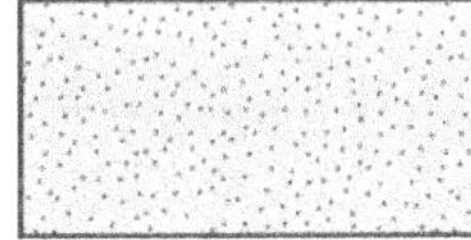

The other option is to place two stones "vertically" adjacent to each other:

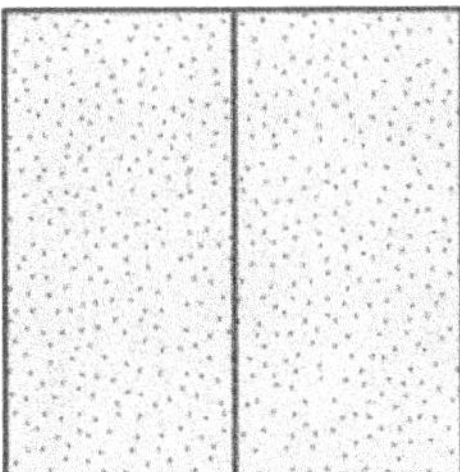

Ask, **In each case, how many more feet of path are needed to build a path that is *n* feet long?** In the first case, the initial part of the path is only 1 foot long, so an additional n – 1 feet are needed. In the second case, the initial part of the path is 2 feet long, so only an additional n – 2 feet are needed.

Ask, **How many ways are there to complete each possible 'beginning'?** Help students to see that the number of n x 2 paths that begin with a horizontal paving stone is a_{n-1}, because the remainder of such a path is simply an (n – 1) x 2 path. Similarly, the number of n x 2 paths that begin with a pair of vertical paving stones at the top is a_{n-2}, because the remainder of the path is simply an $(n - 2) \times 2$ path.

For example, of the eight patterns shown earlier for a 5 x 2 path, there are five with a horizontal paving stone at the top, and three with a pair of vertical paving stones at the top.

Some students may develop the recursive formula from this general analysis, rather than from the numerical data. That's fine; there is no right or wrong order in which to think about this.

The Closed Form

There is a closed-form expression for the number of $n \times 2$ paths, but finding this expression requires knowing (or guessing) that it should be the sum of two exponential expressions. The number of $n \times 2$ paths is given by the expression

$$\frac{1}{\sqrt{5}}\left[\left(\frac{1+\sqrt{5}}{2}\right)^{n+1} - \left(\frac{1-\sqrt{5}}{2}\right)^{n+1}\right]$$

Amazingly, this gives a positive integer for every positive integer value of n. To get the nth term of the Fibonacci sequence, simply replace $n + 1$ with n.

Key Questions

What is this type of formula called?

What does this formula say about the paving patterns?

Why does this recursive formula hold true?

What can the 'beginning' of the paving pattern look like?

In each case, how many more feet of path are needed to build a path that is *n* feet long?

How many ways are there to complete each possible 'beginning'?

More Beach Adventures

Intent

In this activity, students continue to work with the new extended definition of the sine function.

Mathematics

As students manipulate the equation 20 sin $(29t) = -14$, they will find that $29t = \sin^{-1}\left(\frac{-14}{20}\right)$, but solving that for t yields a solution that is not within a meaningful range relative to the problem situation. This introduces the concept of the *principal value* of the inverse sine function. Students will use their knowledge of the periodic nature of the sine function to find relevant values for t using the principal value for the inverse sine.

Progression

The first problem of this activity uses the situation from *Sand Castles*. While many students will have settled for approximate graphical solutions to that activity, they should now be encouraged to find more exact solutions to the question in the current activity. Question 2 gives the students further practice with using the principal value for the inverse sine and the periodicity of the sine function.

Approximate Time

30 to 35 minutes for activity (at home or in class)

10 to 15 minutes for discussion

Classroom Organization

Individuals, followed by whole-class discussion

Doing the Activity

You may want to briefly review the use of the inverse sine function for this activity.

Discussing and Debriefing the Activity

Question 1 is somewhat similar to Question 4 of *Sand Castles.* You may want to give students a few minutes to compare ideas on Question 1, and then have a student explain his or her solution.

The problem involves solving the equation $w(t) = -14$, which means finding values of t that fit the condition 20 sin $(29t) = -14$. But students need to consider the periodicity of the function in order to find the two solutions for t that correspond to the evening hours. For instance, if they consider "evening" to mean a time between 6 p.m. and midnight, then they need t to be a value between 18 and 24.

By using the inverse sine function, students will come up with the "basic" solution to the equation. The equation $29t = \sin^{-1}\left(\frac{-14}{20}\right)$ yields $t \approx -1.53$, which corresponds to about an hour and a half before midnight *on the previous day.* There are several ways that students might find a solution for the evening of the day in question.

One approach is to use the condition $29t \approx -44.4$ to get other values for the expression $29t$. Using the graph of the function $w(t)$ (shown here) may help in understanding what's happening.

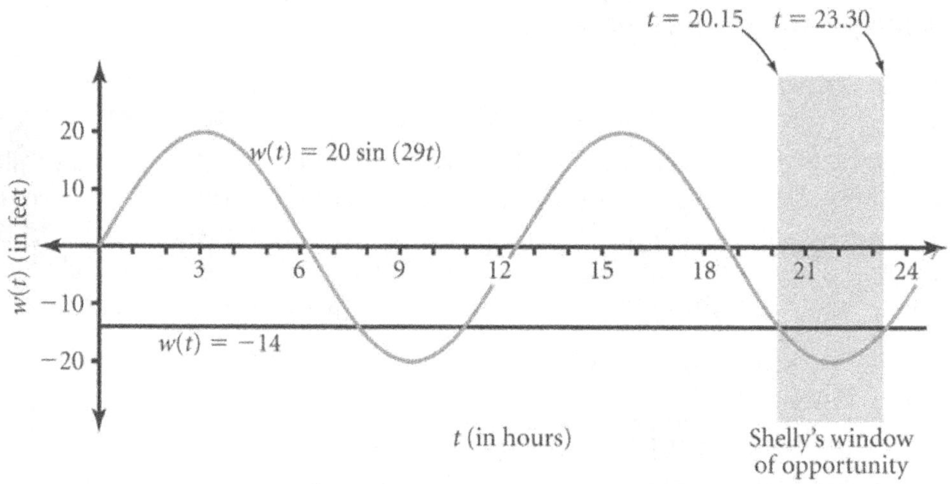

Given that the basic solution is $29t \approx -44.4$, students can use the graph to see that Oceana and her friend can pass by where the rocks jut into the water between $29t = 540° + 44.4°$ and $29t = 720° - 44.4°$. This gives the time interval from $t \approx 20.15$ and $t = 23.30$, which means from about 8:09 p.m. to about 11:18 p.m.

Question 2

Questions 2a through 2d involve a more abstract version of the issues in Question 1, although these questions are simpler because they involve merely sin θ instead of sin $(29t)$. You might have several volunteers give an answer for each of the questions. Some students may point out that each question has not just three but four angles that work. If so, you could ask whether every given value of the sine will produce four angles on this interval, and have students justify their responses using their "plain" sine graph.

Principal Value of the Inverse Sine Function

For Question 2d, many students will have used the "$\sin^{-1}$" key on their calculators. Point out that the calculator has to somehow choose one of many answers as its output for $\sin^{-1}(-0.71)$, and so it is important to standardize the process for doing this.

You can let students explore for a few minutes to see how the calculator's "$\sin^{-1}$" key works. They should discover these facts:

- If x is between 0 and 1, $\sin^{-1} x$ is between 0° and 90°.
- If x is between 0 and –1, $\sin^{-1} x$ is between 0° and –90°.

Of course, if the calculator is given a number with absolute value greater than 1, it gives an error message when you use $\sin^{-1}$; see the next subsection, "The Domain of Inverse Sine."

Note: With most scientific calculators, you enter the number (–0.71, for instance) and then push the "$\sin^{-1}$" key. With most graphing calculators, you do the reverse.

Tell students that although each of the answers to Question 2d is *an* inverse sine for –0.71, the value that the calculator gives (approximately –45.2°) is called the *principal value* for the inverse sine of –0.71. We write this using notation such as $\sin^{-1}(-0.71) \approx -45.2°$. Bring out that the general definition of the inverse sine involves a *convention* for choosing these principal values.

Note: Some textbooks introduce the notation "Sin^{-1}" (with a capital "S") to represent the principal value, and use "$\sin^{-1}$" to be "multivalued." This is not standard mathematical notation, however.

You may want to point out the analogy between this distinction and a similar distinction with square roots. For example, the number 9 has two square roots, 3 and –3, but the symbol $\sqrt{9}$ represents only the number 3, which is sometimes called the *principal square root* of 9.

The Domain of Inverse Sine

Ask, **What's $\sin^{-1}(2)$?** Have students find it on their calculators. They should get some kind of error message.

Ask students to explain this. As a hint, ask them to express the problem in terms of the sine function (rather than in terms of inverse sine). They should be able to describe this as looking for a solution to the equation sin $x = 2$. You may want to let them discuss this equation in their groups until someone can explain why this problem has no solution.

Use this example to review terminology by asking, **What is the domain of the inverse sine function? What is its range?** Bring out that the domain is the interval from −1 to 1, because the equation sin $x = c$ has a solution only if $-1 \leq c \leq 1$ or, in other words, because the range of the sine function is the interval from −1 to 1.

The range of $\sin^{-1}$ is the set of angles from −90° to 90°, inclusive, although this is the result of the conventions for principal values.

Key Questions

What's $\sin^{-1}$ (2)?

What is the domain of the inverse sine function? What is its range?

Supplemental Activity

Prisoner Revisited (reinforcement) is similar to this activity and can be assigned if students need further practice working with the ideas from this activity.

Falling, Falling, Falling

Intent

In this section, students consider the falling motion of the diver in the central unit problem.

Mathematics

Analysis of the falling motion of the diver is complicated by the fact that he accelerates as he falls. The activities in this section use an area model for distance to help students develop general formulas for freely falling bodies. But the analysis is further complicated by the fact that in order to apply the formula for falling time to the unit problem, it must be expressed in terms of the diver's height when he begins his fall, which depends in turn on the time he spends on the Ferris wheel. So students must merge their formula for the time it takes a body to fall with the one they developed earlier for the height of the platform.

Progression

In *Distance with Changing Speed* and *Acceleration Variations and a Sine Summary,* students learn that they can find the average speed for an interval of constant acceleration by averaging the speeds at the beginning and ending of the interval. They apply this principle to the constant acceleration of falling bodies in *Free Fall*, developing formulas for the height of a falling body at any point in time and for the time it will take a body to fall. In *A Practice Jump*, students express this latter formula in terms of the time the diver spends on the moving Ferris wheel platform.

Distance with Changing Speed

Acceleration Variations and a Sine Summary

Free Fall

Not So Spectacular

A Practice Jump

Distance with Changing Speed

Intent

In this activity, students develop a method for finding the total distance traveled in situations involving constant acceleration.

Mathematics

In the introduction to this activity, students see that they can express total distance traveled in terms of area under the graph of the speed function. This leads to a discovery that under constant acceleration the average speed for a time interval is the average of the initial and final speeds for that interval.

Progression

Introduce this activity with a brief discussion about using an area model to represent distance in terms of speed. Students apply this concept as they work on the activity individually, then share their results in class discussion.

Approximate Time

5 to 10 minutes for introduction

20 to 25 minutes for activity (at home or in class)

10 to 20 minutes for discussion

Classroom Organization

Individuals, followed by whole-class discussion

Materials

Transparency of *Distance with Changing Speed* blackline master

Doing the Activity

Remind students that one type of motion they will be considering will be that of the falling diver, whose speed changes as he falls. Tell them that because the speed is changing, the relationship among the variables of distance, speed, and time is more complex than if the speed were constant. Inform them that in order to understand this complex situation, they will develop a simple model, using a graph, for representing the distance a moving object (or person) travels in terms of its speed.

Begin by posing this straightforward question:

Suppose a person drives for 3 hours at a constant speed of 50 miles per hour. How far does the person go?

All students need to do here is multiply the speed (50 miles per hour) by the time (3 hours) to get the distance (150 miles).

Have students make a graph showing speed as a function of time for this situation. Their graphs should look like this:

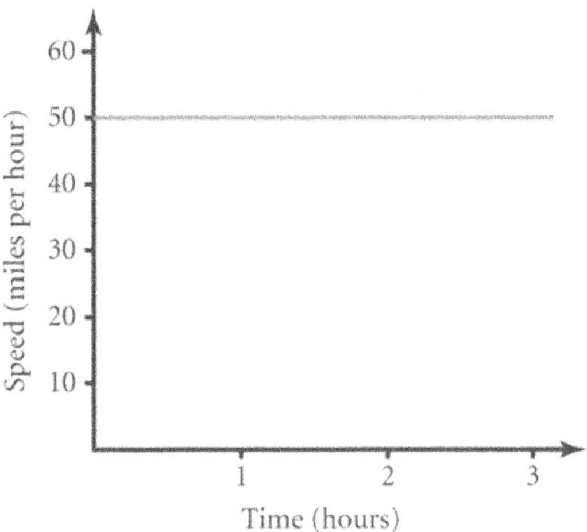

Then ask, **How can you represent the distance geometrically?** As a hint, remind students that for constant speed, distance is simply the product of the speed and the time. As a further hint, remind them that the area of a rectangle is often a good model for multiplication.

These hints should lead the class to form a rectangle as in the next diagram and to see that the area of this rectangle gives the numerical value of the distance traveled. You may want to suggest that students subdivide the rectangle, as shown here, to indicate the distance covered each hour.

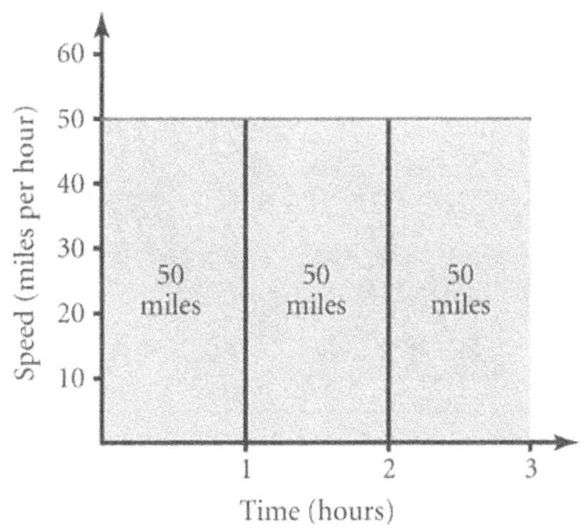

Tell the class that this activity continues the use of this area model for finding total distance.

Comment: It may seem strange to use an area model to represent a linear measurement, as the discussion here does. But because *distance* is the product of *rate* and *time*, it is appropriate to use a two-dimensional model for distance. (You may want to acknowledge this anomaly to students.)

With the preceding introduction to the idea of an area model for finding distance, have groups begin work on the activity.

Discussing and Debriefing the Activity

Question 1 extends the use of the area model beyond the case of constant speed. For Question 1a, students should get a diagram something like this, showing speed as a function of time:

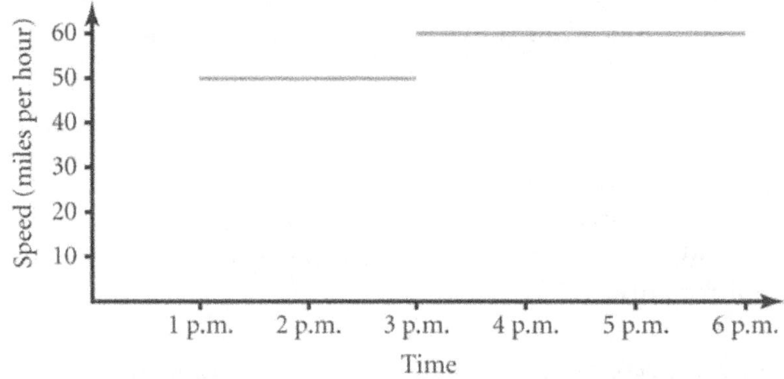

For Question 1b, students might draw in the rectangles, as shown in the next diagram, to illustrate that the distance traveled from 1 p.m. to 3 p.m. is the area of the first rectangle and the distance traveled from 3 p.m. to 6 p.m. is the area of the second rectangle. Thus, the total area under the graph is equal to the total distance traveled.

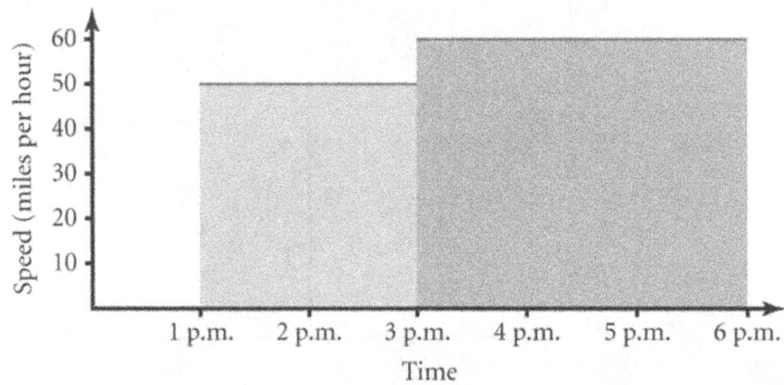

Question 2

Question 2 applies the area model to a situation in which speed is changing at a constant rate. For Question 2a, students should get a graph like the following. (You may want to use a transparency of this graph, provided in the *Distance with Changing Speed* blackline master, to aid in the discussion.)

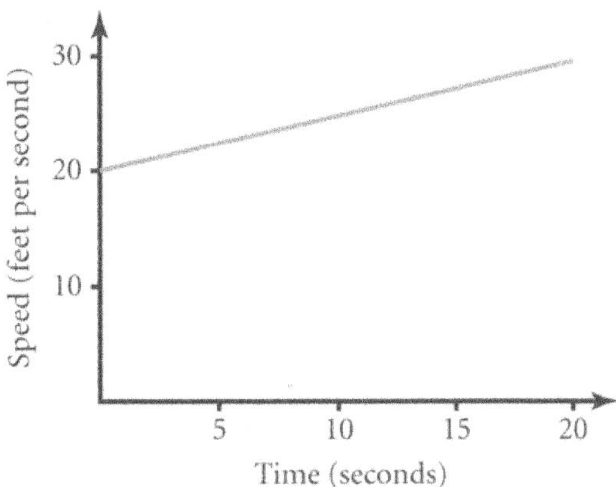

For Question 2b, students will likely take a purely intuitive approach, saying that because the speed increases at a constant rate from 20 feet per second to 30 feet per second, the average speed is simply 25 feet per second.

The main goal of this activity is to confirm this intuitive approach by using the area model. Based on the earlier examples, students should accept that the total distance traveled ought to be equal to the area under the graph, as shown here:

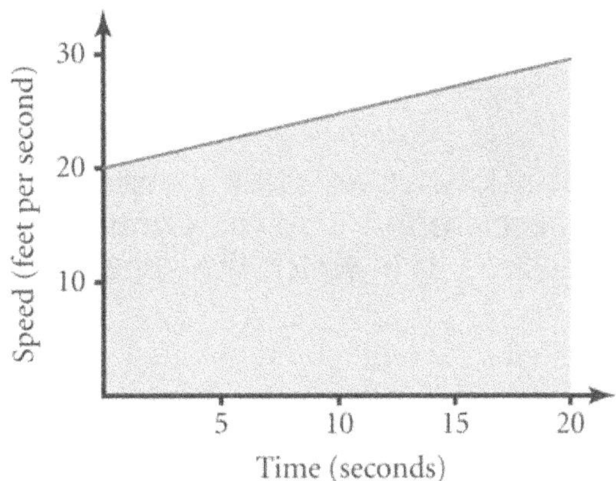

Ask, **How do you find the area of this shaded figure?** Students may recall that the figure is called a trapezoid, and they might even recall the formula for the area of a trapezoid. But don't get sidetracked by area formulas here. One intuitive approach is to ask, **What rectangle would have the same area, using the same base?** Students should see that the rectangle shown in this diagram has the same area as the trapezoid:

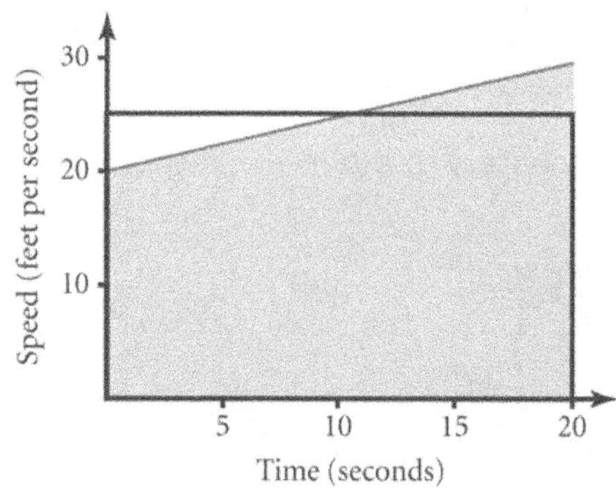

Students should also be able to see that the height of this rectangle is 25—which is the average of the heights of the two ends of the trapezoid—and that the area of the rectangle, and hence of the trapezoid, is 500. You may want to have them express the area in the form $\left(\frac{20+30}{2}\right) \cdot 20$ and review that this is an illustration of the general formula for the area of a trapezoid.

Ask, **What does the area mean in terms of the runner?** Bring out that it means that the runner goes a total of 500 feet.

Finally, verify that this result is consistent with the intuitive answer to Question 2b. That is, if the runner averaged 25 feet per second, as in Question 2b, and went at this speed for 20 seconds, he would travel 500 feet, as found in Question 2c using area.

Ask, **What do we call the rate at which speed is changing?** (If no one comes up with the term, you might ask what we call it when a car speeds up.) If no one knows, introduce the term *acceleration.* Use the phrase *constant acceleration* to describe the situation in Question 2, in which the speed is changing at a constant rate.

Averaging the Endpoints

Bring out that the trapezoid approach will work for any situation of constant acceleration and that it gives a simple way to find the total distance traveled, even though the speed is not constant.

Ask students to summarize the principle for finding average speed in situations of constant acceleration. They should be able to articulate something like this, which you should post:

If an object is traveling with constant acceleration, then its average speed over any time interval is the average of its beginning speed and its final speed during that time interval.

We will refer to this principle as the "averaging the endpoints" method for finding average speed. Emphasize to students that it applies only to situations with constant acceleration.

You might also ask how this principle can be used to find the total distance traveled. Students should see that, as always, they can multiply the average speed by the length of the time interval to get the total distance.

Key Questions

How can you represent the distance geometrically?

How do you find the area of this shaded figure?

What rectangle would have the same area, using the same base?

What does the area mean in terms of the runner?

What do we call the rate at which speed is changing?

Supplemental Activity

Lightning at the Beach on Jupiter (reinforcement) is a review of the relationship between distance, rate, and time.

Acceleration Variations and a Sine Summary

Intent

In this activity, students continue to work with the area model for distance, and they also summarize ideas about extending the sine function.

Mathematics

The first part of this activity looks at how the principle of averaging the beginning and ending speeds to find the average speed of a time interval is affected if we abandon the requirement that the acceleration be constant. The second part asks students to summarize their work in extending the sine function and to connect this new definition with the unit problem. The main purpose of Part II is to get them thinking about the issues as preparation for a class discussion.

Progression

Students work on the activity individually and share their results in class discussion.

Approximate Time

30 minutes for activity (at home or in class)

10 minutes for discussion

Classroom Organization

Individuals, followed by whole-class discussion

Doing the Activity

You may want to give some students transparencies to prepare graphs for the discussion of Part I of this activity.

Important: If the class has not completed the discussion of *Distance with Changing Speed,* you should postpone Part I of this activity.

Discussing and Debriefing the Activity

Part I: Acceleration Variations

Let a different volunteer share an example for each of the variations called for in Part I of the activity. You might give presenters transparencies, and have them duplicate the axes and graph shown in the activity and then sketch their graphs along with it. Like the graph in the activity, the graphs students create should show the speed going from 20 feet per second to 30 feet per second. Have each presenter explain how he or she knows that the graph fits the given condition.

More about Acceleration

Point out that the units for acceleration are somewhat complicated. Speed itself is measured here in feet per second, and acceleration measures how speed changes over time. In the example given in the problem, the runner's speed increases by 10 feet per second over a 20-second interval, so it increases by 0.5 feet per second for each second that elapses. Tell students that we express this by saying that the acceleration is 0.5 feet per second per second.

Emphasize the distinction between *speed,* which in this case tells the rate at which the runner's *position* is changing, and *acceleration,* which tells how fast the runner's *speed* is changing. Here, the *speed* is not constant (the runner's *position* is not changing at a constant rate), but the *acceleration* is constant (the runner's *speed* is changing at a constant rate).

Part II: A Sine Summary

Let students share their ideas with the class about the extension of the sine function and how it fits with the problem. Students should feel fairly comfortable at this stage of the unit with the idea that the expression 65 + 50 sin (9*t*) gives the height of the platform after *t* seconds (starting from the 3 o'clock position).

Free Fall

Intent

In this activity, students examine the behavior of falling objects.

Mathematics

In the discussion that introduces this activity, students recognize that falling objects accelerate, and they learn that their acceleration is constant. They then use the "averaging the endpoints" method to find formulas for the height of an object falling from rest in terms of time, and for the time it takes for an object to fall h feet.

Progression

This activity is introduced with a brief discussion of falling objects. Students then work in groups, and discuss their results as a class.

Approximate Time

10 minutes for introduction

30 minutes for activity

25 minutes for discussion

Classroom Organization

Small groups, followed by whole-class discussion

Doing the Activity

Falling Objects

Quickly review the principle of "averaging the endpoints." Then ask, **What does 'averaging the endpoints' have to do with the main unit problem?** If students don't see a connection, ask, **What happens to an object as it falls freely?** Let students share their own ideas and experiences about what happens as objects fall.

If no one has a convincing argument that falling objects gain speed, then you may want to bring up a situation like this for them to consider:

Which would hurt more, a fall from your roof or a fall from your bed?

(Although this example relates more directly to the force of impact than to speed, most students will attribute the added force of falling from a roof to moving faster at impact.)

Tell students that although our experience shows us that objects go faster and faster as they fall, physicists actually know precisely how falling objects behave. Specifically, from experimental data and from theoretical considerations, they know that falling objects have *constant acceleration.* You may want to post this statement, perhaps adjacent to the description of the "averaging the endpoints" method:

> **Falling objects have constant acceleration (under ideal circumstances). That is, the speed of a falling object changes at a constant rate.**

Discuss that the phrase "under ideal circumstances" means that there is no wind, air resistance, or other complicating factor to interfere with the object's fall. That is, the principle describes the behavior of *free-falling* objects. (This assumption is mentioned in the activity.) You might also discuss the fact that this assumption is reasonable for some types of objects (such as rocks) and not for others (such as feathers).

Free Fall

Have students read the introduction to the activity *Free Fall* and go over the details in the section "Starting from Rest." Then have them work on the questions. If necessary, suggest that for Question 1, they use the "averaging the endpoints" method.

Note: Question 5 explicitly states that the diver falls "from rest." If the issue of the effect of the Ferris wheel's motion has come up before (see the subsection "For Teachers: The Diver's Initial Motion" in the discussion notes for *The Circus Act*), you may want to remind students that they are assuming for now, in their work on the central unit problem, that the diver falls from rest. See the subsection "But There's More to the Problem!" in the discussion notes for *Moving Cart, Turning Ferris Wheel*.

Discussing and Debriefing the Activity

Let one or two students present their analysis for Question 1. Question 1a should be straightforward. That is, if students understood the introduction to the activity, they should see that the speed at $t = 5$ is simply $5 \cdot 32 = 160$ ft/s.

To find out how far an object falls in 5 seconds, students should reason that the object's average speed for that interval is equal to the average of its instantaneous speeds at the endpoints of the interval. These endpoints are $t = 0$ and $t = 5$. The information in the activity tells them that the instantaneous speed at $t = 0$ is 0 ft/s and the instantaneous speed at $t = 5$ is 160 ft/s.

The average of 0 and 160 is 80, so the object has an average speed over the 5-second interval of 80 ft/s. Therefore, the object falls $5 \cdot 80 = 400$ feet during this interval.

Question 2

The key element in this activity is for students to generalize the reasoning from Question 1 to develop the general formula asked for in Question 2.

The approach we expect is for students to see that the instantaneous speed at the end of t seconds is $32t$ and the instantaneous speed at the start is 0. Thus, the average speed over the first t seconds is $16t$ ft/s. Students can then multiply this average speed by the length of the time interval, which is t seconds, to get a total distance traveled of $16t^2$ feet.

Post this conclusion, because it will play a critical role throughout the rest of the unit:

> **If an object falls freely from rest, it will fall $16t^2$ feet in its first t seconds.**

Note: Physicists often use the term "displacement" and the letter s for the distance an object has traveled. Some students may have seen the formula $s = 16t^2$ in a physics class to describe the displacement of an object falling from rest.

Questions 3 and 4

For Question 3, students simply need to subtract $16t^2$ from the initial height h to get the expression $h - 16t^2$ for the object's height after t seconds. You can add this additional conclusion to the statement just posted:

> **If the object's initial height is h feet, then its height after t seconds is $h - 16t^2$ feet.**

Question 4 may seem straightforward once the formula $h - 16t^2$ has been found, but be sure to go over the transition carefully. For some students, it may be a substantial step from the idea of "reaching the ground" to the step of setting $h - 16t^2$ equal to 0. Help them to understand that answering Question 4 is equivalent to solving the equation $h - 16t^2 = 0$ for t in terms of h. Have a volunteer show the details of solving the equation to get $t = \sqrt{\frac{h}{16}}$ seconds.

Once students have developed this additional generalization, you should post it along with the previous formula. This new statement might say something like this:

If an object falls freely from rest, it will take $\sqrt{\frac{h}{16}}$ seconds for the object to fall *h* feet.

Question 5

Question 5 provides an important variation, in which students need to see that the diver is actually falling 82 feet. They might set this up with the equation $90 - 16t^2 = 8$, or they might simply set $16t^2$ equal to 82. In either case, they should get the expression $\sqrt{\frac{82}{16}}$, which means that it takes approximately 2.26 seconds for the diver to fall to the water level.)

The Number 32 Is an Approximation

At some point, bring out that the number 32, which appears in *Free Fall*, is a numerical approximation based on experiments. (The activity does say "approximately" in giving the value of 32 feet per second for each second of the object's fall, but students may overlook this.)

Also point out that this number is specific to the use of feet as the unit of length. For example, if we instead measure length in meters, then we use approximately 9.8 instead of 32 (because 32 feet is about 9.8 meters).

Key Questions

What does 'averaging the endpoints' have to do with the main unit problem?

What happens to an object as it falls freely?

Supplemental Activity

In *The Derivative of Position* (extension), students use derivatives to confirm one of the formulas they developed in this activity.

Not So Spectacular

Intent

In this activity, students again work with the fact that there are many angles with the same sine.

Mathematics

In this activity, students develop a general expression for the times at which the diver will be at a given height.

Progression

This activity resembles *A Clear View,* but deals with the periodicity of the Ferris wheel's motion. This activity also puts some of the ideas from *More Beach Adventures* into the context of the Ferris wheel.

Approximate Time

30 minutes for activity (at home or in class)

25 minutes for discussion

Classroom Organization

Individuals, followed by whole-class discussion

Doing the Activity

Students work on this activity independently.

Discussing and Debriefing the Activity

Give students a few minutes to compare ideas, and perhaps select a student at random to present a solution for the activity. Up to a point, this problem is essentially the same as *A Clear View*. In both problems, students must find the points in the cycle of the Ferris wheel where a person is a given height off the ground. The difference in this problem is the focus on obtaining the general solution. Use this discussion to review the ideas about the inverse sine function and principal values discussed with *More Beach Adventures.*

Initial Solutions

The diagram here shows the two positions at which the owner would be 25 feet off the ground and gives details about one of these locations. This shows that the angle

θ must satisfy the condition $\sin\theta = \frac{40}{50}$, so $\theta = \sin^{-1} 0.8$, which is approximately 53.1°. This means that the large angle shown as $9t$ is approximately equal to 360° - 53.1° = 306.9°. In other words, the Ferris wheel could have been turning for about 306.9 ÷ 9 = 34.1 seconds. Another approach is to find $\sin^{-1}(-0.8) = -53.1°$, then add 360°, leading to the same result.

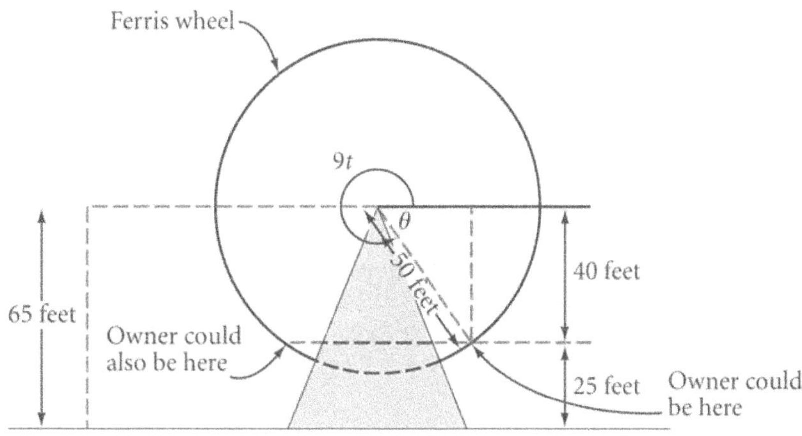

For the other time at a height of 25 feet, we have

$$9t \approx 180° + 53.1° = 233.1°,$$

which means $t \approx 233.1 \div 9$ or 25.9 seconds.

To this point, the problem is basically a repetition of the approach in *A Clear View*.

Generalizing the Solutions

The new aspect of this problem is that each of these two locations actually leads to an infinite list of possible answers to the question of how long the Ferris wheel has been turning, because the Ferris wheel may have gone around several times before reaching the given point. (If students did not realize that the activity called for this additional stage, let them work on it in groups, trying to find the general solutions.)

Two main approaches are likely to come about. Here are descriptions of these two approaches for the position at the lower right of the diagram:

- Find different possibilities for the total angle $9t$ by adding multiples of 360° to the basic angle of 306.9°, and then divide each total angle by 9 (degrees per second).
- Find the time of 34.1 seconds for the basic angle of 306.9° and then add multiples of 40 seconds to that time.

In either case, one gets 34.1 seconds, 74.1 seconds, 114.1 seconds, and so on, as the sequence of possible times. Students should be able, perhaps with some prodding, to express this general solution as something like 34.1 + 40n, where n can be any positive integer.

For the position at the lower left of the diagram, the same steps apply, but the general solution is 25.9 + 40n seconds.

Bring out that all of these values satisfy the equation $\sin(9t) = -0.8$, but none of them correspond directly to the principal value of the inverse sine function, which gives a negative value for t. Ask whether the two expressions together provide *all* possible solutions to $\sin(9t) = -0.8$. Help students see that this is true if n is allowed to be any *integer*, but that the solutions generated by negative values of n (e.g., –5.9 seconds, – 45.9 seconds, or –14.1 seconds, – 54.1 seconds, and so on) do not fit the physical realities of this problem.

A Practice Jump

Intent

In this activity, students combine principles about falling objects with the general sine function in connection with the unit problem.

Mathematics

This activity asks students to combine their new formula for falling time with their earlier work on the height of the platform at a given position, and to generalize this in terms of the amount of time the Ferris wheel has been turning.

Progression

Students combine the principles from several previous activities as they work on this activity. They work individually, and then share their results in class discussion.

Approximate Time

25 to 30 minutes for activity (at home or in class)

15 to 25 minutes for discussion

Classroom Organization

Individuals, followed by whole-class discussion

Doing the Activity

Students find the time it will take the platform to reach the 11 o'clock position, then find the diver's height at that position, and finally find the time it will take him to fall from that height. Students then generalize this to a formula for the diver's falling time for a release time of W seconds after the Ferris wheel starts turning.

Discussing and Debriefing the Activity

Presentations of this activity will be a good indicator of how well students understood their work in previous activities. Let different volunteers present different parts of the problem. As usual, ask for alternate approaches after the presentations.

Questions 1 and 2

One way to find the "let go" time (Question 1) is for students to see that the 11 o'clock position is one-third $\left(\frac{4}{12}\right)$ of the way around, so it should take one-third of the time of a full rotation to get there. Another approach is to find the angle of rotation from the 3 o'clock position to the 11 o'clock position (120°) and divide this by the angular speed of 9 degrees per second.

In either case, students should see that the assistant should let go after about 13.33 seconds (or that value plus any multiple of 40 seconds).

To solve Question 2, students can substitute the time just found, about 13.33 seconds, into the general height formula, $h = 65 + 50 \sin (9t)$. This will give them the height of 108.3 feet. They can also use the angle directly, getting the height from the expression $65 + 50 \sin 120°$.

Question 3

For Question 3, students need to subtract 8 feet from the height of 108.3 feet (found in Question 2) to get the distance the diver will fall before he hits the water in the cart. That is, the diver falls 100.3 feet.

To complete Question 3, students can substitute the value $h = 100.3$ into the expression $\sqrt{\frac{h}{16}}$. They should find that the diver's falling time is about 2.50 seconds.

Question 4

For Question 4, students need to combine the general formula for the diver's height after t seconds on the Ferris wheel with the formula from *Free Fall* for the time required to fall h feet. Essentially, this is merely substitution, but it is difficult for some students to make this transition. Go slowly, and be sure that presenters explain themselves clearly.

You may want to refer repeatedly to the specifics of Question 3 as this generalization is developed. Here are the key steps of the generalization:

- After W seconds, the diver is at a height of $65 + 50 \sin (9W)$ feet.
- The diver needs to fall 8 feet less than this, for a total falling distance of $57 + 50 \sin (9W)$ feet.
- The time it takes an object at rest to fall $57 + 50 \sin (9W)$ feet is $\sqrt{\frac{57 + 50 \sin (9W)}{16}}$ seconds.

Falling Time is a Function of Turning Time

Bring out that students are expressing the time the diver spends in the air in terms of the amount of time the Ferris wheel turns before the assistant lets go. You might do this through a series of questions like these:

- First ask, **What does the diver's falling time depend on?** Students will probably say that it depends on his height when the assistant lets go.
- Then ask, **What does this height depend on?** Students will probably say it depends on what the diver's position is when the assistant lets go.
- Finally, ask, **What does this position depend on?** They should see that it depends on how long the Ferris wheel is turning before the assistant lets go.

Students may shortcut one or more of these steps. For example, they may say that the falling time depends on where the diver is, or even that it depends on how long the Ferris wheel is turning. The important thing is that they trace the falling time back to the amount of time the Ferris wheel turns before the assistant lets go.

Then ask, **What are you trying to find in the main unit problem?** They should see that they are being asked to find out how long the assistant should let the Ferris wheel turn before letting go.

In other words, what they are looking for is what was called *W* in Question 4 of the activity and in the formula they found for that question. That is, they have just found a formula for the diver's falling time in terms of the variable that they are trying to find in the unit problem.

This can be summarized like this:

If the Ferris wheel passes the 3 o'clock position at $t = 0$ and turns for W seconds more before the assistant lets go, then the diver will fall for

$$\sqrt{\frac{57 + 50 \sin(9W)}{16}}$$

seconds before reaching the level of the water in the cart.

Post this conclusion prominently. It synthesizes two major ideas—the time required for a free-falling object to fall a given distance and the diver's height at the time of release—into a single important formula. You may want to add it on to the poster for the height of the platform after *W* seconds (see the discussion notes for *Testing the Definition*).

Because this expression for falling time is so complex, you may find it helpful to use a single letter to represent it. We will use F ("falling time"), so F is given by the formula

$$F = \sqrt{\frac{57 + 50 \sin(9W)}{16}}$$

If you use this abbreviation, be sure students keep in mind that F is a function of W. You may want to have several students express in their own words what F represents.

Key Questions

What does the diver's falling time depend on?

What does this height depend on?

What does this position depend on?

What are you trying to find in the main unit problem?

Moving Left and Right

Intent

In this section, students focus on the horizontal dimension of the central unit problem.

Mathematics

In the previous sections of this unit, students explored the vertical dimension of the unit problem. They have looked at the vertical position of the platform as the Ferris wheel turns and they have examined the diver's falling motion after he leaves the platform. Now they will focus on horizontal motion.

As the Ferris wheel rotates, the horizontal position of the platform changes, too. Determining the x-coordinate of the platform will involve the cosine function, so, as was done previously with the sine function, the definition of the cosine function will need to be expanded to allow for angles beyond the first quadrant.

While the falling motion of the diver has only a vertical component in this simplified version of the unit problem, the length of time that the cart moves horizontally is in part determined by the time it takes the diver to fall. So an analysis of the cart's motion will include elements of both the horizontal and vertical movement of the platform, as well as of the diver's fall.

Progression

The cart begins to move at the same instant that the Ferris wheel begins to turn, so its motion spans both the time that the diver is on the moving platform and the time that the diver is falling. Students fit these facts together into an expression for the cart's travel time in *Cart Travel Time*. They then return to consideration of the horizontal position of the diver on the platform, first looking at specific times in *Where Does He Land?*, then developing a general formula in *First Quadrant Platform*, and finally using that formula as a basis for extending the cosine function to arbitrary angles in *Generalizing the Platform*. Students return to building a formula for the cart's position in *Carts and Periodic Problems*. In *Planning for Formulas* they consolidate all of the formulas they have developed in preparation for solving the unit problem.

Cart Travel Time

Where Does He Land?

First Quadrant Platform

Carts and Periodic Problems

Generalizing the Platform

Planning for Formulas

Cart Travel Time

Intent

In this activity, students begin to explore the horizontal dimension of the unit problem.

Mathematics

The solution to the unit problem requires that the cart and the diver end up in the same place at the same time. So far, students have considered the movement of the diver. Now they will turn their attention to the movement of the cart. In this activity, they develop an expression for the time that the cart will travel, including both the time that the diver is on the platform and the time that the diver is falling from the platform to the water level.

Progression

Students work in groups to write an expression in terms of W for the time that the cart will travel, from the moment it starts until the moment the diver reaches the level of the water. In the subsequent discussion, you'll introduce the horizontal axis of the Ferris wheel coordinate system.

Approximate Time

15 to 20 minutes for activity

10 to 15 minutes for discussion and introduction of horizontal axis

Classroom Organization

Small groups, followed by whole-class discussion

Doing the Activity

In this activity, students use their conclusion about how long the diver is falling to reach a conclusion about the amount of time the *cart* is traveling. The key is to look at the cart's travel time in two parts:

- The time between when the platform passes the 3 o'clock position and when the diver is released from the platform
- The time while the diver is falling from the platform

In a later activity, *Carts and Periodic Problems,* students will find the cart's *position* at the moment when the diver reaches the water level.

Discussing and Debriefing the Activity

Only a brief presentation should be needed here, because students already have expressions for each of the two parts of the cart's travel time.

- The cart travels for W seconds from when the platform passes the 3 o'clock position until the diver is released.
- The cart travels for F seconds while the diver is falling (where F is given in terms of W by the expression $F = \sqrt{\frac{57 + 50 \sin(9W)}{16}}$).

Post a summary of the conclusion from this activity:

If the cart begins moving when the Ferris wheel passes the 3 o'clock position, and the diver is released *W* seconds later, then the cart will travel for *W* + *F* seconds before the diver reaches the level of the water in the cart, where *F* is given by the formula

$$\boldsymbol{F = \sqrt{\frac{57 + 50 \sin(9W)}{16}}}$$

The Horizontal Dimension

Bring out that most of students' work on the unit problem has focused on vertical motion—the height of the platform as it turns and the falling motion of the diver.

Ask, What will determine the success or failure of the circus act? The cart needs to be in the right place along its horizontal path when the diver reaches the water level. So point out to students that they must explore the horizontal dimension as well.

Introduce the horizontal axis as shown in the diagram below. The center of the base of the Ferris wheel represents zero, positions to the right of the Ferris wheel are considered to have positive x-coordinates, and distances are measured in feet. Tell students that we will identify an object's horizontal position within this system using its x-coordinate.

To clarify this system, ask, **What is the cart's x-coordinate at the start of the circus act?** They should see that the cart's initial x-coordinate is −240, because the cart is 240 feet to the left of the center of the base of the Ferris wheel.

Similarly, ask, **What about the platform?** Students should see that the platform's initial x-coordinate is 50, because the platform starts at the 3 o'clock position and the radius of the Ferris wheel is 50 feet.

Key Questions

What will determine the success or failure of the circus act?

What is the cart's x-coordinate at the start of the circus act?

What about the platform?

Where Does He Land?

Intent

In this activity, students find the platform's x-coordinate for specific cases.

Mathematics

This activity is like a combination of *As the Ferris Wheel Turns* and *Graphing the Ferris Wheel,* except that it deals with horizontal instead of vertical position. Students use the cosine function in different quadrants to find the diver's x-coordinate when he falls and to graph the x-coordinate as a function of time.

Progression

Students work individually as they explore the diver's horizontal position. Students will extend this activity to develop a general formula in *First Quadrant Platform*.

Approximate Time

30 minutes for activity (at home or in class)

15 to 20 minutes for discussion

Classroom Organization

Individuals, followed by whole-class discussion

Materials

Transparency of the *Where Does He Land?* blackline master

Doing the Activity

Question 1 asks students to find the diver's x-coordinate at each of five given release times. Question 2 asks them to graph the x-coordinate as a function of t.

Discussing and Debriefing the Activity

This unit has adopted the informal convention of using t as a general time variable and W as the value of t when the diver is released. But both t and W are often used to represent the same quantity: the number of seconds the Ferris wheel turns before release. Clarify this if students express confusion over the use of these two variables.

For Question 1, let several students present their results for specific values of t, using diagrams to explain their answers. As these presentations are discussed, bring out that the x-coordinate of the diver's landing position is the same as the x-coordinate of the platform at the moment the diver is released.

For instance, for $t = 3$ (Question 1a), the Ferris wheel has turned 27°, and the presenter might give a diagram like this:

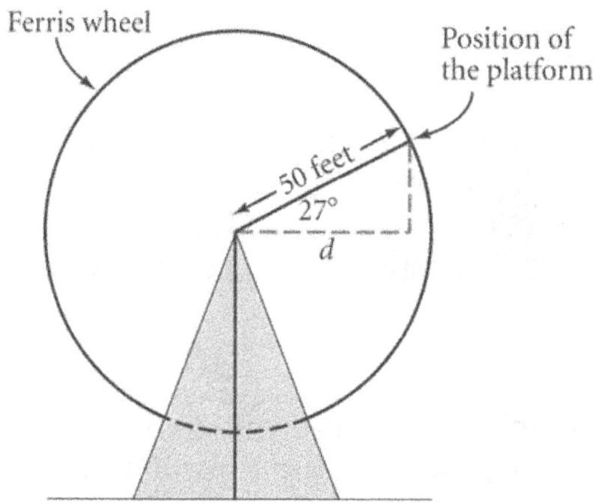

This gives $\cos 27° = \frac{d}{50}$, so the platform's x-coordinate is 50 cos 27°, or approximately 44.55. This means that while the diver is falling, his x-coordinate is also approximately 44.55. (Students may simply say that the diver lands about 44.55 feet to the right of the center of the Ferris wheel. Help them make the transition from this description to the use of the coordinate terminology.)

For the case $t = 12$ (Question 1c), the platform is in the second quadrant when the diver is released, and students are likely to express their common x-coordinate as −50 cos 72°, as illustrated in the next diagram. It is important to bring out that although the segment labeled d has length 50 cos 72° (because $\cos 72° = \frac{d}{50}$), the x-coordinate must be negative. That is, $x = -50 \cos 72°$.

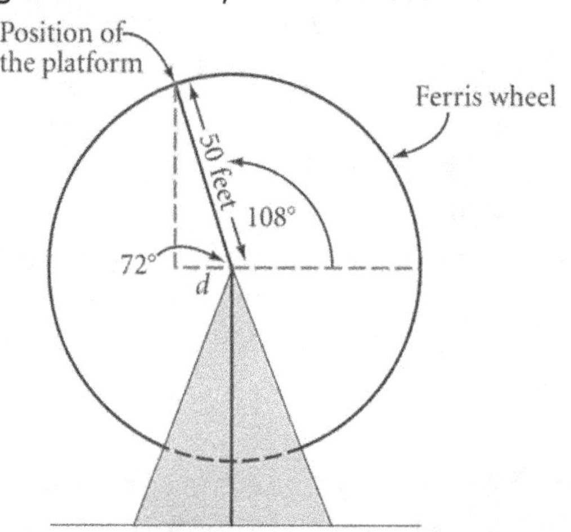

Similarly, for $t = 26$, students are likely to get $x = -50 \cos 54°$, while for $t = 37$, they will probably get $x = 50 \cos 27°$. (Students might express these answers using the sine function instead, but ultimately, the goal is to get a general formula in terms of cosine.)

Question 2

You can use a transparency of the blank coordinate system for this activity, provided in the *Where Does He Land?* blackline master, or you can have students develop the scales for the axes themselves. In either case, you or the students can plot individual points as they are suggested. Be sure to get a variety of points from $t = 0$ through $t = 80$. The graph should look roughly like this:

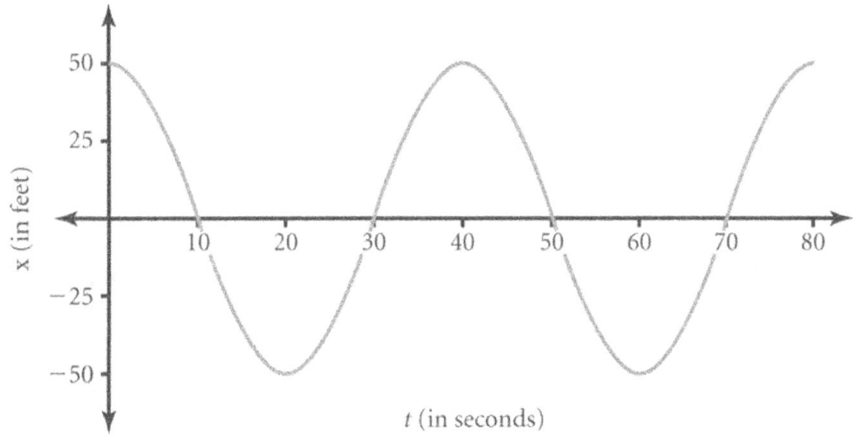

There are two main observations to make about this graph:

- This graph makes sense in terms of the Ferris wheel problem. For example, it is periodic with period 40, and it shows the platform being farthest to the right at $t = 0$, 40, and 80, farthest to the left at $t = 20$ and 60, and having a zero x-coordinate at $t = 10$, 30, 50, and 70.
- This graph is similar to the graph of vertical position that students made in *Graphing the Ferris Wheel*, with two key differences:
    a) This graph has its maximum at $t = 0$ seconds, while the graph of the platform's height had its maximum at
    $t = 10$ seconds.
    b) This graph is "balanced" around the horizontal axis, while the graph of the platform's height was "balanced" around the height of 65 feet.

First-Quadrant Platform

Intent

In this activity, students generalize the platform's x-coordinate for the first quadrant, in preparation for extending the cosine function definition to one that is meaningful for arbitrary angles.

Mathematics

In *Where Does He Land?* students calculated the platform's x-coordinate at specific points in time. Now they generalize this to develop a formula for the platform's x-coordinate in the first quadrant. They are not yet able to extend this to the other quadrants because the definition of the cosine that they are working with is based upon right-triangle geometry and therefore is only meaningful for acute angles. Students will develop a more general definition of the cosine function for arbitrary angles in *Generalizing the Platform*.

Progression

First-Quadrant Platform asks students to generalize their work from *Where Does He Land?* to find a formula for the position of the platform in terms of t in the first quadrant. This activity is very similar to *At Certain Points in Time* and should require little discussion.

Approximate Time

15 to 25 minutes for activity (at home or in class)

5 minutes for discussion

Classroom Organization

Individuals, followed by whole-class discussion

Doing the Activity

In this activity, students develop a general formula for the platform's x-coordinate for the first quadrant (as they did for the platform's height in *At Certain Points in Time*). They will then use that formula to motivate the general definition of the cosine function. No introduction is needed, and probably only a brief discussion will be required.

Discussing and Debriefing the Activity

Let a student present his or her result. The class should be able to see that the platform's x-coordinate is given by the equation $x = 50 \cos (9t)$.

Carts and Periodic Problems

Intent

In this activity, students find a formula for the cart's position and consider examples of periodic behavior.

Mathematics

Students combine their earlier formula for the cart's travel time with the newly introduced horizontal coordinate system to find a formula for the position of the cart. They also think further about situations that involve periodic motion.

Progression

Students work on the activity individually or in groups, then discuss their results as a class.

Approximate Time

25 minutes for activity (at home or in class)

10 to 15 minutes for discussion

Classroom Organization

Individuals or small groups, followed by whole-class discussion

Doing the Activity

In *Cart Travel Time*, students found an expression for the cart travel time in terms of W, the time that the Ferris wheel has been turning. Now they combine that with the information about the speed and starting coordinate of the cart to come up with a formula for the cart's x-coordinate when the diver reaches water level.

Part II asks students to describe, state the period, and sketch graphs of several situations that they believe are periodic.

Discussing and Debriefing the Activity

Part I: Where's the Cart?

Have a volunteer present Part I. In *Cart Travel Time*, students saw that from $t = 0$ until the diver reaches the water level, the cart travels for $W + F$ seconds (where F represents the diver's falling time and is given by the expression $\sqrt{\frac{57 + 50 \sin(9W)}{16}}$). Students need to combine that information with the facts about the cart's speed and initial position to find that the cart's x-coordinate when the diver reaches the water level is $-240 + 15(W + F)$.

Post this conclusion:

> **Suppose the diver is dropped after W seconds on the Ferris wheel (starting from the 3 o'clock position). When the diver reaches the water level, the cart's x-coordinate is**

$$\mathbf{-240 + 15(W + F) \text{ where } F = \sqrt{\frac{57 + 50\sin(9W)}{16}}}.$$

Part II: Periodic Problems

You can give overheads to a couple of groups and have them choose one or two of their most interesting examples to share with the class. They should present the situation, give the period, and show a sketch of the graph.

Here are some of the many ideas they might mention:

- Phases of the moon
- Menstrual cycles
- The movement of the hands of a clock
- The height of the sun in the sky

Generalizing the Platform

Intent

In this activity, students extend the cosine function to be defined for all angles.

Mathematics

Similar to what students saw in *Extending the Sine*, they are here asked to redefine the cosine in a way that makes sense for arbitrary angles.

The subsequent discussion emphasizes that there is only one way to extend the cosine function that will allow the first-quadrant formula for the platform's x-coordinate to work in all quadrants. After the cosine function has been formally defined, students graph the function. The discussion connects the cosine function to the Ferris wheel problem, bringing out that students now have general formulas for the diver's vertical and horizontal positions as functions of time.

Progression

Students work in groups to define the cosine function beyond the first quadrant. In the discussion, they graph this function, and connect it to the Ferris wheel problem.

Approximate Time

20 to 30 minutes for activity

15 to 25 minutes for discussion

Classroom Organization

Small groups, followed by whole-class discussion

Doing the Activity

Tell students that this activity is basically a cosine version of earlier work with the sine function. To extending the cosine function beyond the right-triangle definition, students seek to define the cosine function for non-acute angles so that the equation $x = 50 \cos (9t)$ will give the platform's horizontal position for all values of t.

Discussing and Debriefing the Activity

Have a student present Question 1. Discussion of this problem should give you a good sense of how well students understand the process they used for generalizing the sine function (in *Extending the Sine* and *Testing the Definition*). They should see that the platform's x-coordinate is $-50 \cos 72°$ for $t = 12$. For the equation $x = 50 \cos (9t)$ to give this value when $t = 12$, they need to define $\cos 108°$ to be equal to $-\cos 72°$.

Use your judgment about whether you need to discuss Question 2 as well.

Question 3

Let another student present Question 3, and provide additional hints and review as needed. The goal is to have students see that in a diagram such as the one shown here, with θ a first-quadrant angle, $\cos \theta$ is equal to $\frac{x}{r}$, and that using this ratio as the general definition for arbitrary angles is consistent with their work in Questions 1 and 2. (If needed, you can go through details such as those used in *Extending the Sine* for the sine function.)

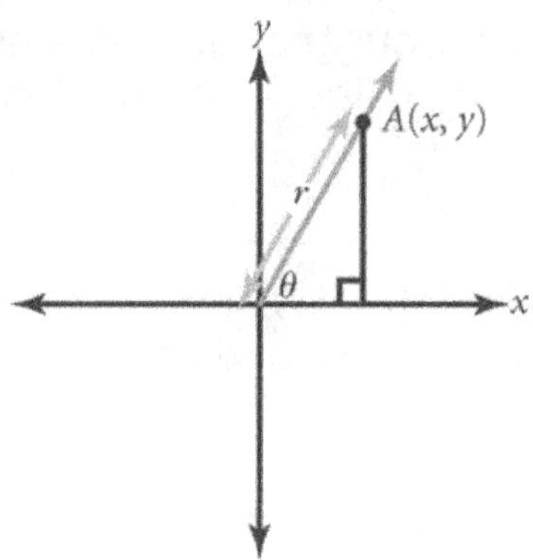

Defining the Cosine Function

When students seem clear about the process, you can simply tell them that this is the basis for the formal definition of the cosine function for arbitrary angles. You should post this together with an appropriate diagram (such as the one that follows):

> **For any angle θ, we define $\cos \theta$ by first drawing the ray that makes a counterclockwise angle θ with the positive x-axis and choosing a point A on this ray (other than the origin) with coordinates (x, y).**
>
> **Using the shorthand $r = \sqrt{x^2 + y^2}$, we then define the cosine function by the equation**
>
> $$\cos \theta = \frac{x}{r}$$

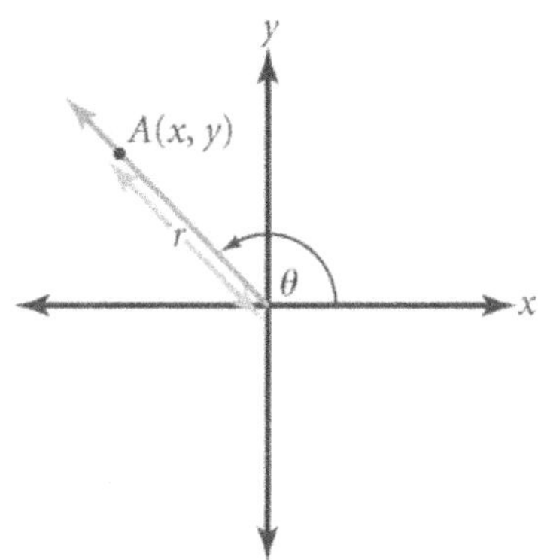

As with the discussion of the sine function, emphasize how well this works, particularly in terms of sign: the extended cosine function is positive when x is positive and negative when x is negative. These two cases correspond exactly to the sign of the platform's x-coordinate, which is positive to the right of center and negative to the left of center.

The Cosine Function Is Well Defined

Briefly point out that the ratio $\frac{x}{r}$ does not depend on the specific point chosen on the defining ray. You can discuss the fact that this can be proved using similarity, just as was done for the sine function.

Cosine Graphs

Have students graph the function defined by the expression 50 cos ($9t$) on their calculators, adjusting the viewing window to include all values from $t = 0$ to $t = 80$. Have them compare the result to their graphs from Question 2 of *Where Does He Land?* They should see that the graphs are the same.

Also have students graph the "plain" cosine function on their calculators, including negative values for the angle, and compare it with the graph of the "plain" sine function. They should see that the two graphs are identical in shape, but one is "shifted" from the other. You can post and label a graph like the one shown here near the graph of the sine function for future reference.

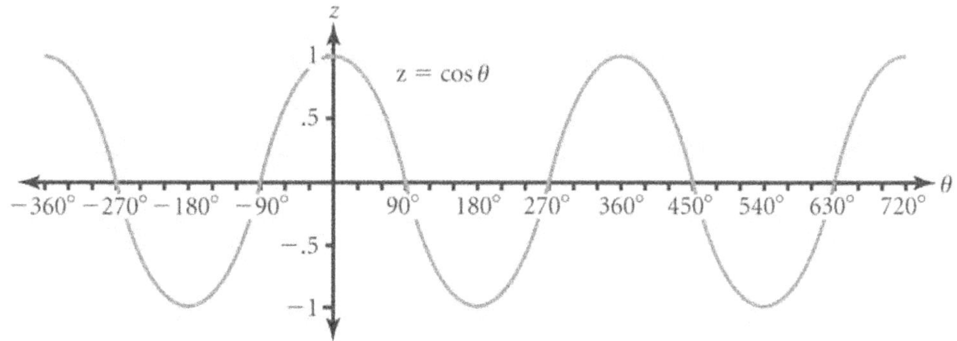

Cosine and the Ferris Wheel

Ask students, **What does this general definition of the cosine function mean in terms of the *x*-coordinate of the platform as the Ferris wheel turns?** They should be able to articulate a conclusion like this, which you should post:

> **If a Ferris wheel of radius 50 feet makes a complete turn every 40 seconds, starting from the 3 o'clock position, then the *x*-coordinate of the platform, after *t* seconds, is given by the function**
>
> $$x = 50 \cos (9t)$$

The Diver's Position When He Reaches the Water Level

Ask how this conclusion fits into the solution of the unit problem. Bring out that (at least for the current simplified version of the problem), the diver's *x*-coordinate as he falls is the same as the *x*-coordinate of the platform when the diver is dropped. In other words:

If the diver is dropped after the Ferris wheel has been turning for W seconds, starting from the 3 o'clock position, then his x-coordinate as he falls is given by the function

$$x = 50 \cos (9W)$$

Key Question

What does this general definition of the cosine function mean in terms of the x-coordinate of the platform as the Ferris wheel turns?

Planning for Formulas

Intent

In this activity, students sum up and explain their work thus far in the unit.

Mathematics

In preparation for fitting all of the pieces together to solve the unit problem in *Moving Cart, Turning Ferris Wheel*, in this activity students review each of the formulas that they have developed in the unit. They explain how the facts of the unit problem are reflected in each formula and what each piece of the formula represents.

Progression

Students work on this activity individually. The discussion reviews each of the formulas in depth, but not how they fit together.

Approximate Time

30 to 40 minutes for activity (at home or in class)

30 to 40 minutes for discussion

Classroom Organization

Individuals, followed by whole-class discussion

Doing the Activity

Students work on this activity independently.

Discussing and Debriefing the Activity

Have several students present elements of each formula. For each formula, you might have one student give the formula itself and have others explain how the facts about the cart and Ferris wheel, the definitions of the sine and cosine functions, and the principles about falling objects fit in. This discussion involves a thorough review of the unit so far, and so it is likely to take a significant amount of time.

Finding the Release Time

Intent

In this section, students solve the central unit problem.

Mathematics

Students now put together the various pieces of the solution that they have accumulated throughout the unit to write an equation whose solution will answer the unit problem. Since the equation cannot be solved through algebraic manipulation, they must also consider what other avenues they have for approximating a solution. Each of the several primary strategies for solving the equation highlights the power of the graphing calculator.

Progression

Students solve the unit problem in *Moving Cart, Turning Ferris Wheel. Putting the Cart Before the Ferris Wheel* helps them to reason through the pieces that were necessary to complete the former activity as they solve a variation on the unit problem. They establish firmly their understanding that the diver and the cart must be in the same place at the same time.

The remaining activities in this section complete the development of the cosine function. Students graph the function and examine what happens in various quadrants in *What's Your Cosine?* They look again at an application of the cosine function in *Find the Ferris Wheel*, as they examine the connection between the parameters of the Ferris wheel and the coefficients in the formula for the *x*-coordinate of the platform.

Moving Cart, Turning Ferris Wheel

Putting the Cart Before the Ferris Wheel

What's Your Cosine?

Find the Ferris Wheel

Moving Cart, Turning Ferris Wheel

Intent

In this activity, students solve the central unit problem.

Mathematics

In *Carts and Periodic Problems*, students developed an expression for the position of the cart when the diver reaches water level in terms of W, the time at which the diver is dropped. In *Generalizing the Platform*, students arrived at a definition of the cosine function that would allow them to apply their formula for the x-coordinate of the platform (and therefore of the diver) to all four quadrants. The key to creating an equation for the unit problem is recognizing that the diver and the cart must both be in the same place when the diver reaches water level.

Setting the expressions from these two activities equal to each other creates an equation that will solve the unit problem, but that equation can't be solved by algebraic manipulation. It can be solved by an organized guess-and-check approach or, more elegantly, by using the graphing calculator to approximate the intersection of the functions from the two earlier activities.

Progression

Students work in groups to solve the unit problem. They will need plenty of time both for formulating the equation, and for thinking through how to solve it. Class discussion will be particularly important for sharing ideas about how to solve the equation, and looking at the issue of just how accurate the answer needs to be.

Approximate Time

50 to 60 minutes for activity

30 minutes for discussion

Classroom Organization

Small groups, followed by whole-class discussion

Doing the Activity

No introduction is needed for this activity.

This is the culmination of a long effort to solve the unit problem. It is important that you allow students sufficient time to work through this problem. Although this activity is "nothing more than" a combination of previous ideas, the process of bringing these varied ideas together may be difficult for many students.

Although some groups may need to start by guessing specific numbers, you should help them move toward the formulation of a specific equation that they can use to find *W*.

Hints as Students Work

As students work on this, you can remind them to focus on the idea of having the cart and the diver be in the same place when the diver reaches the water level. You can also suggest that they look around the room (or in their notes, or at their work on *Planning for Formulas*) for summaries of key elements of the analysis.

There are several forms for an equation that states that the cart is in the right position when the diver reaches the water level. This is one possibility:

$$-240 + 15\left(W + \sqrt{\frac{57 + 50 \sin(9W)}{16}}\right) = 50 \cos(9W)$$

Once students get this or an equivalent equation, they will need some time to come to the realization that algebraic manipulation is not going to work. The discussion below suggests several approaches for getting a numerical estimate of the solution.

Discussing and Debriefing the Activity

Let several students present their analyses of the problem. Give further hints as needed to get an equation that expresses the fact that the cart is in the right place. Again, here is one possibility:

$$-240 + 15\left(W + \sqrt{\frac{57 + 50 \sin(9W)}{16}}\right) = 50 \cos(9W)$$

Make sure students can articulate that the left side of the equation gives the *x*-coordinate of the cart at the time the diver is at water level, and the right side of the equation gives the *x*-coordinate of the diver at that same time. In other words, the equation is saying that the diver is landing in the water.

Students may come up with variations on this equation. Let students present any alternate approaches on the overhead, and ask them to articulate their meaning.

Help students appreciate that this analysis is "quadrant-free." That is, it works no matter where the platform is when the diver is released. You might remind them that in the initial activities of the unit (such as *As the Ferris Wheel Turns*), they did not have the general sine and cosine functions, and their analysis was slightly different for each quadrant.

Solving the Equation

Elicit presentations by groups that have solved the equation by different methods, including guess-and-check and graphing. Here are three ways students might solve the complicated equation just developed:

- Guess-and-check: That is, pick a value for W, evaluate both sides, and then repeatedly adjust W to bring the two sides of the equation closer together.
- Graphing: For instance, graph the two functions defined by the expressions on the two sides of the equation and then look at where the graphs meet. This will require adjusting the window settings in order to locate the point of intersection.
- Using a "solve" feature on a calculator.

This may be a good time to discuss how to approach entry of complicated equations into the calculator (perhaps by breaking them into smaller pieces) and how to use the "solve" feature on the calculator.

It turns out that the assistant should release the diver about 12.28 seconds after the cart starts moving.

Ta-Da!

Give a cheer! The problem is solved! If students haven't already done so, you will probably want to have them substitute $W = 12.28$ into both sides of the equation to confirm that this value really is correct.

Also, be sure to answer the question posed in *Moving Cart, Turning Ferris Wheel* of where the diver is on the Ferris wheel when the assistant releases him. After 12.28 seconds, the Ferris wheel will have turned $9 \cdot 12.28 \approx 110.5°$, which will place the platform between the 11 o'clock and 12 o'clock positions.

The diver's height off the ground when released is given by the expression $65 + 50 \sin (9W)$, which gives a value of about 112 feet. The diver's x-coordinate is given by the expression $50 \cos (9W)$, which comes out to about -17.5, which means he is about 17.5 feet to the left of center.

Students will probably want to work out some more of the stages in the process for $W = 12.28$. For example, the diver must fall about $112 - 8 = 104$ feet, which will take about $\sqrt{\frac{104}{16}} \approx 2.55$ seconds. (That is, $F = 2.55$.) Thus, the cart must travel a total of about 12.28 + 2.55 = 14.83 seconds.

How Much Accuracy Is Needed?

Because the value 12.28 is an approximate solution to a practical problem, you should raise the question, **How accurate does the assistant need to be?** For instance, if he drops the diver after 12.3 seconds, what will happen?

Students can explore this simply by substituting other values for *W* into the two sides of the equation to see how much effect changes in *W* have on the *x*-coordinates of the cart and the diver. For instance, if $W = 12.3$ seconds, then at the moment the diver reaches the water level, the cart's *x*-coordinate is approximately −17.3 and the diver's *x*-coordinate is approximately −17.7. As long as the tub is of a reasonable size, these few inches should not matter. On the other hand, for $W = 12.4$, the cart's *x*-coordinate is approximately −15.9 and the diver's *x*-coordinate is approximately −18.4, so this represents a difference of several feet, which might be problematic.

But There's More to the Problem!

As noted earlier, the problem as solved just now involves a simplification—assuming that the diver falls as if from rest. The more complex version of the problem is the subject of the Year 4 unit *The Diver Returns.*

Some suggestions are provided in the discussion notes for *"High Dive" Portfolio* for encouraging students to speculate on the more complex version of the problem, although you might have them do this now.

Key Question

How accurate does the assistant need to be?

Putting the Cart Before the Ferris Wheel

Intent

In this activity, students look at a question that is similar to the central unit problem but much simpler.

Mathematics

Working on this simpler problem may help some students with *Moving Cart, Turning Ferris Wheel.* It will particularly assist them in realizing that the key to the unit problem is to ensure that the cart and the diver are in the same place when the diver reaches water level.

Progression

Students work on the activity individually and then discuss their results as a class.

Approximate Time

25 to 35 minutes for activity (at home or in class)

10 to 15 minutes for discussion

Classroom Organization

Individuals, followed by whole-class discussion

Doing the Activity

Putting the Cart Before the Ferris Wheel asks where the cart would need to start in order to catch the diver if he is released 25 seconds after the Ferris wheel began turning.

Discussing and Debriefing the Activity

Let volunteers explain each of Questions 1 and 2. Question 1 should be a review of familiar ideas. The diver's x-coordinate as he falls (and when he lands) is $50 \cos (9 \cdot 25)$, or approximately -35.4.

Question 2

Students need to put several ideas together to answer Question 2. First, they need to find out how long the diver is in the air. If he is released after 25 seconds on the Ferris wheel, his height off the ground (in feet) will be $65 + 50 \sin (9 \cdot 25)$, or approximately 29.6 feet, and so he will fall 21.6 feet to reach the water level. This means that his falling time (in seconds) will be $\sqrt{\frac{21.6}{16}}$, or approximately 1.16 seconds.

Once they have found this falling time, students can think about the cart's movement. The cart will travel a total of approximately 26.16 seconds (the initial 25 seconds plus 1.16 seconds for the diver's falling time). At 15 feet per second, this means the cart will travel approximately 392.4 feet. Therefore, to be in the right place when the diver reaches the water level, the cart must start 392.4 feet to the left of −35.4, which means the cart must start 427.8 feet to the left of center.

What's Your Cosine?

Intent

In this activity, students graph the cosine function and consider its periodicity.

Mathematics

This activity gives students experience with the new extended definition of the cosine function. The questions reinforce the vocabulary associated with periodic functions, make students think about the sign of the cosine in each quadrant, and give them practice working with the symmetry of the graph.

Progression

Students work on the activity individually and then discuss their results as a class.

Approximate Time

30 minutes for activity (at home or in class)

15 to 20 minutes for discussion

Classroom Organization

Individuals, followed by whole-class discussion

Doing the Activity

Question 1 asks students to draw a graph of the cosine function and to name its amplitude, period, intercepts, and angles for which it has maxima and minima. Question 2 has students find multiple angles that have the same or opposite cosine as a given angle.

The subsequent discussion includes consideration of the sign of the cosine function in each quadrant.

Discussing and Debriefing the Activity

Let volunteers each answer a component of the activity. The discussion of Question 1 can be similar to that for *The "Plain" Sine Graph.*

Question 2

By this time students should be able to argue with confidence why each part of Question 2 has four possible answers. Use the discussion of Question 2 to get students to generalize about the sign of the cosine function. By looking at the sign of the x-coordinate, they should be able to determine that the cosine function is positive if the angle is in the first or fourth quadrant, and negative if the angle is in the second or third quadrant. You may find it helpful to use a diagram like this:

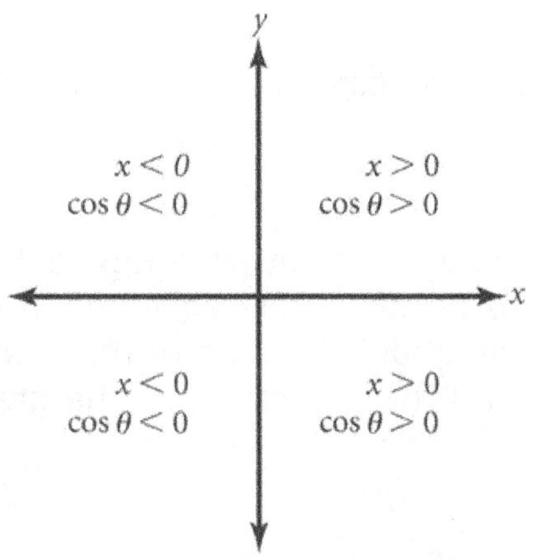

Find the Ferris Wheel

Intent

This activity offers an opportunity to confirm that students understand the connections between the parameters of the Ferris wheel problem and the coefficients in the formula for the x-coordinate of a rider on the Ferris wheel.

Mathematics

In this activity, students examine the connection between the function describing a rider's horizontal position and the parameters of the Ferris wheel. They also look at the effect of changes in the function on its graph.

Progression

Students work on the activity individually and share their results in class discussion.

Approximate Time

15 to 25 minutes for activity (at home or in class)

10 minutes for discussion

Classroom Organization

Individuals, followed by whole-class discussion

Doing the Activity

Question 1 presents two variations of the equation for the x-coordinate of the platform on the Ferris wheel and asks students to explain what values for the physical parameters of the Ferris wheel are represented by each equation. Question 2 asks students to write a similar expression for a Ferris wheel with a smaller radius but greater angular speed than that in Question 1a, and to explain how the graphs of the two expressions would differ.

Discussing and Debriefing the Activity

There may be some confusion over period versus angular speed. The latter is the number that appears in the equation. For example, in Question 1a, the coefficient 10 in the expression $25 \cos(10t)$ means that the Ferris wheel turns 10 degrees per second. To get the period, students need to divide 360° by this coefficient (or do some equivalent arithmetic process).

Question 2

Students will probably all have different answers for Question 2, and you can use two or three of these answers to illustrate the options. Be sure to get at least a verbal description of how the graph for such a function would differ from that in Question 1a.

After discussion of Question 2, use your judgment about whether to take the time to have students actually graph examples involving the variations described. If students seem able to articulate that a smaller radius will lead to a smaller amplitude of the graph and that a faster angular speed will lead to a more "scrunched up" graph, then they probably have a sufficient understanding of the ideas.

A Trigonometric Conclusion

Intent

In this section, students study additional trigonometric topics.

Mathematics

While students have solved the central unit problem at this point, the Ferris wheel situation is a useful context for introducing several more concepts in trigonometry. Describing the diver's position on the Ferris wheel platform in terms of the angle of rotation lends itself nicely to the introduction of polar coordinates, and the analogy is equally helpful in developing several trigonometric identities that can be visualized using the symmetry of the Ferris wheel. This section also extends the definition of the tangent function to all angles, as was already done for the sine and cosine functions.

Progression

Polar coordinates are introduced in *Some Polar Practice* and *A Polar Summary*, and students gain further insights into this topic through their work in *Polar Coordinates on the Ferris Wheel.*

Students develop an important Pythagorean identity in *Pythagorean Trigonometry*. In *Positions on the Ferris Wheel*, they discover that the sines of supplementary angles are equal, and in *More Positions on the Ferris Wheel* students consider the sine and cosine of a negative angle.

Coordinate Tangents extends the tangent function to arbitrary angles. Students summarize what they have learned in this unit in *A Trigonometric Reflection* and *"High Dive" Portfolio*.

Some Polar Practice

A Polar Summary

Polar Coordinates on the Ferris Wheel

Pythagorean Trigonometry

Coordinate Tangents

Positions on the Ferris Wheel

More Positions on the Ferris Wheel

A Trigonometric Reflection

"High Dive" Portfolio

Some Polar Practice

Intent

This activity is an introduction to polar coordinates.

Mathematics

The Ferris wheel situation lends itself to consideration using polar coordinates. In this activity, students are introduced to polar coordinates, and they find rectangular coordinates from polar coordinates and vice versa. This will be connected to the Ferris wheel in *Polar Coordinates on the Ferris Wheel*.

Progression

Begin this activity by introducing polar coordinates, bringing out that a point has many representations in polar coordinates. Students will then work on the activity in groups. The subsequent discussion develops general equations for expressing rectangular coordinates in terms of polar coordinates, and includes an introduction to the use of negative angles, angles greater than 360°, and negative values for r in polar coordinates.

Approximate Time

10 to 15 minutes for introduction

20 to 30 minutes for activity

10 minutes for discussion

Classroom Organization

Small groups, followed by whole-class discussion

Doing the Activity

Polar Coordinates

Ask, **How have you been describing the platform's** (or diver's) **position on the Ferris wheel?** Emphasize that students have described this position in terms of the platform's height off the ground and its horizontal position relative to the center of the Ferris wheel.

Ask, **Where is the origin of the coordinate system?** Help students articulate that they have been working with an *xy*-coordinate system whose origin is at ground level, directly below the center of the Ferris wheel.

Then ask, **What information have you been using to get these coordinates?** Bring out that both coordinates are expressed in terms of the radius of the Ferris wheel and the angle through which the platform has turned. Tell them that because the turning occurs at the center of the Ferris wheel, rather than at ground level, it makes more sense to treat the center of the Ferris wheel itself as the origin.

Next, inform students that in the coordinate plane, the measurements corresponding to the radius of the Ferris wheel and the angle of turn are called the **polar coordinates** of a point and are usually represented by the variables r and θ. (We suggest that you wait until after *Some Polar Practice* before introducing the fact that points in the plane have many polar representations. See the subsection "Multiple Answers" in the discussion below.)

Illustrate the idea of polar coordinates with a diagram as shown here, clarifying the role of each variable:

- The polar coordinate r represents the distance from point A to the origin.
- The polar coordinate θ represents the counterclockwise angle made between the positive direction of the x-axis and the ray from the origin through point A.

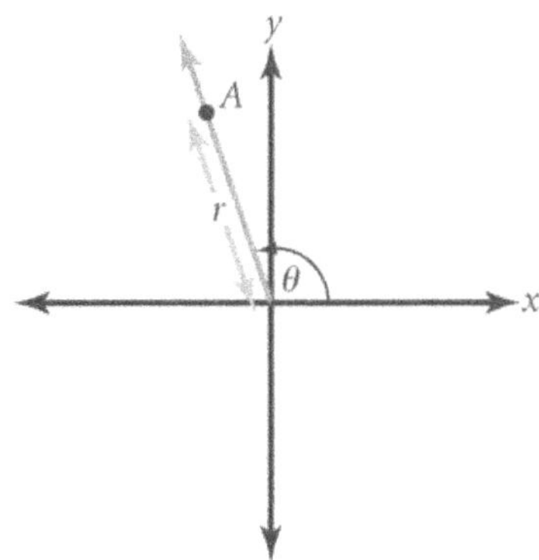

Tell students that in giving the polar coordinates of a point, we give the r value first, by convention.Illustrate this with an example. For instance, ask, **What are the polar coordinates for a rider on the 50-foot Ferris wheel at the 11 o'clock position?** Students should see that this position would be represented as (50, 120°).

Point out that with two systems of coordinates under discussion, it's important to be clear which system is being used. Review the term *rectangular coordinates* as another name for the system of x- and y-coordinates.

Some Polar Practice

With the preceding introduction of the basic definitions, have groups work on *Some Polar Practice.*

Discussing and Debriefing the Activity

Let several students present individual problems. Have the class discuss whether the answers seem reasonable. For instance, on Question 1a, students should see that both the *x*-coordinate and the *y*-coordinate should be positive (because the point is in the first quadrant) and that the *x*-coordinate should be larger than the *y*-coordinate (because the angle is less than 45°). On Question 1b, they should see that the *x*-coordinate is negative and the *y*-coordinate is positive because the point is in the second quadrant.

To the nearest hundredth, the answers are

- For Question 1a: (1.73, 1)
- For Question 1b: (–3.83, 3.21)

The General Equations

Ask, **In general, how can you find the rectangular coordinates of a point from its polar coordinates?**

If students need a hint, you might suggest that they look at the general definitions of the sine and cosine functions. Students should see that they can simply multiply each defining equation by r. That is, $\cos\theta = \frac{x}{r}$ (by definition), so $x = r\cos\theta$. Similarly, $y = r\sin\theta$. Point out that these relationships work in all quadrants, and post these equations.

Post this result:

If a point has polar coordinates r and θ, then its rectangular coordinates x and y can be found by the equations

$$x = r\cos\theta \quad \text{and} \quad y = r\sin\theta$$

Question 2

For Question 2, students are likely to give only the "obvious" answers for the two examples, which are approximately (8.25, 14°) for Question 2a and (9.85, 294°) for Question 2b. (If they give other answers, that will lead smoothly into the next subsection.)

Multiple Answers

Tell students that by convention, we allow *any* angle, not merely those from 0° to 360°, to be considered as the angle in polar coordinates. Bring out that an angle of more than 360° simply represents more than one complete turn around the origin. Also clarify that negative angles are measured clockwise (from the positive direction of the *x*-axis).

Have students provide at least a couple of alternate solutions for each of Questions 2a and 2b, one with an angle of more than 360° and one with a negative angle. For instance, the answer to Question 2a could be (8.25, 374°) or (8.25, –346°), and the answer to Question 2b could be (9.85, –66°) or (9.85, 654°). Tell students that because a given point has more than one representation, we sometimes speak of "a polar representation" of a point rather than "the polar coordinates" of the point.

You can use the Ferris wheel model to illustrate this idea of multiple representations, pointing out that a rider will pass the same place many times, which gives many different polar coordinate representations of that position. Students will probably see that these representations all have the same *r*-coordinate. But they should realize that there are infinitely many values for θ that correspond to a given point.

Tell students that we also allow negative values for *r*, by going in the opposite direction from the ray defined by θ. Either give an example or elicit one from the class. For instance, help students see that the point in Question 2a could also be represented in polar coordinates by the pair (–8.25, 194°).

Without getting into details, we suggest that you point out that because each point has many polar coordinate representations, there are no simple formulas for getting a polar representation of a point from its rectangular coordinates.

Direct students to the *A Polar Summary* reference pages for a summary of the topics of this discussion.

Key Questions

How have you been describing the platform's position on the Ferris wheel?

Where is the origin of the coordinate system?

What information have you been using to get these coordinates?

What are the polar coordinates for a rider on the 50-foot Ferris wheel at the 11 o'clock position?

What's another name for the system of *x*- and *y*-coordinates?

In general, how can you find the rectangular coordinates of a point from its polar coordinates?

Supplemental Activities

Polar Equations (*extension*) introduces the graphing of polar equations.

Circular Sine (*extension*) challenges students to transform a polar equation into a rectangular equation.

A Polar Exploration (*extension*) is an open-ended exploration of the graphs of polar equations.

A Polar Summary

Intent

A Polar Summary is reference material that introduces polar coordinates.

Mathematics

These reference pages summarize the representation of points in polar coordinates and the conversion of rectangular coordinates to polar coordinates. It includes discussion of polar coordinates with angles greater than 360°, negative angles, and negative values for r.

Progression

The material in this activity is simply a summary of the discussion notes from *Some Polar Practice*. Direct students to this reference material after the discussion following that activity.

Approximate Time

0 to 10 minutes for discussion

Classroom Organization

For reference and/or discussion only

Doing the Activity

After the class has completed and discussed *Some Polar Practice*, refer students to the summary of polar coordinates in *A Polar Summary.*

Discussing and Debriefing the Activity

You may want to review these ideas briefly again, referring to the text of *A Polar Summary*.

Polar Coordinates on the Ferris Wheel

Intent

In this activity, students look at the relationship between polar coordinates and the Ferris wheel problem.

Mathematics

This activity gives students further experience working with polar and rectangular coordinates while enabling them to tie their new knowledge of the polar coordinate system to the familiar context of the Ferris wheel. This context helps to reinforce the concept that every point in the coordinate plane has many ways to be represented in polar coordinates.

Progression

Students work on the activity individually and then share their results as a class.

Approximate Time

20 to 30 minutes for activity (at home or in class)

10 minutes for discussion

Classroom Organization

Individuals, followed by whole-class discussion

Doing the Activity

Polar Coordinates on the Ferris Wheel asks the student to find both polar and rectangular coordinates for a rider on a Ferris wheel at a given time, to find a different time at which the rider would be in the same position and the associated polar coordinates, and then to find a general expression for each set of coordinates at time t.

Discussing and Debriefing the Activity

This discussion will show you how well students are making the connection between polar coordinates and the Ferris wheel. For Question 1, they should see that 3 seconds represents a turn of 54° (based on the period of 20 seconds), so the "obvious" polar coordinates are (40, 54°). To get the rectangular coordinates, students can think in terms of either "coordinate formulas" or "Ferris wheel formulas" to see that the rectangular coordinates can be expressed as (40 cos 54°, 40 sin 54°), which is approximately (23.5, 32.4).

Question 2a illustrates the periodicity of the motion, and students might give values such as 23 seconds, 43 seconds, and so on. For Question 2b, they might use angles of 414°, 774°, and so on. For Question 3, the only "work" needed to get the polar coordinates is the computation that each second represents 18° of turn, so the rider's polar coordinates after t seconds are (40, $18t°$). The rectangular coordinates are simply a variation on the formulas students found for the rider's position in the main unit problem.

Pythagorean Trigonometry

Intent

In this activity, students develop a Pythagorean identity.

Mathematics

The Pythagorean identity $\sin^2 \theta + \cos^2 \theta = 1$ is the single most important relationship among the trigonometric functions. This activity will help students discover this relationship.

Progression

Students work in groups to discover the Pythagorean identity. The subsequent discussion introduces the term *identity* to describe a general relationship such as the equation just developed, and reviews the identity $\sin \theta = \cos (90^\circ - \theta)$.

Approximate Time

25 minutes for activity

15 minutes for discussion

Classroom Organization

Small groups, followed by whole-class discussion

Doing the Activity

This activity asks students to combine the coordinates of a point on the unit circle (expressed in terms of x, y, and trigonometric functions) and the equation of the unit circle to write an equation relating $\sin \theta$ and $\cos \theta$. Students should develop the identity $\sin^2 \theta + \cos^2 \theta = 1$. Students then choose an angle in each quadrant to verify that the relationship holds true.

You may want to review the idea that the definitions of sine and cosine are independent of the choice of point along the appropriate ray, and that we sometimes use a point on the unit circle for the definitions.

Discussing and Debriefing the Activity

Let different students present results for each question. For Question 1, presenters will probably substitute 1 for r in $\sin \theta = \frac{y}{r}$ and $\cos \theta = \frac{x}{r}$, and come up with $y = \sin \theta$ and $x = \cos \theta$. For Question 2, they should get $x^2 + y^2 = 1$. On

Question 3, if students replace x and y as intended, they will get the equation $(\cos\theta)^2 + (\sin\theta)^2 = 1$.

A Notation Convention

Tell students that by convention, the square of $\cos\theta$ is written as $\cos^2\theta$ and the square of $\sin\theta$ is written as $\sin^2\theta$ (and similarly for the other trigonometric functions). Thus, we usually write the equation from Question 3 as

$$\cos^2\theta + \sin^2\theta = 1.$$

You may want to check occasionally to verify that students understand that $\cos^2\theta$ really means $(\cos\theta)^2$, and so on. In particular, if students try to use the "$\cos^2\theta$" notation on their calculators, they will discover that the calculator will not accept it. For instance, if they want to find $\cos^2 20°$, they will need to enter the expression $(\cos 20)^2$.

Beyond the Unit Circle

Point out that Questions 1 through 3 are set up using points on the unit circle, and ask, **Does the relationship $\cos^2\theta + \sin^2\theta = 1$ require you to use points on the unit circle?** Students might simply point out that the equation doesn't involve r, so once the equation has been established (using the point for which $r = 1$), the value of r no longer matters.

Question 4

Use your judgment about whether to take class time to go over the verification of the equation for specific values.

The Pythagorean Identity

Tell students that the relationship

$$\cos^2\theta + \sin^2\theta = 1$$

is known as a *Pythagorean identity.* (There are other Pythagorean identities.)

Explain that this equation is called an **identity** because it is true for all values of the variable, that is, for all angles θ. The "Pythagorean" part of the name comes from its connection with the Pythagorean theorem.

Post this relationship with its name and a statement that this equation is true for all values of θ.

A Familiar Trigonometric Identity

Students have previously seen the trigonometric identity sin θ = cos (90° − θ) and may have reviewed it earlier in this unit. But in previous discussions of this relationship, they knew only the right-triangle definitions of the trigonometric functions. In the next part of the discussion, students are reminded of this equation as another illustration of an identity, and find that it is true for all angles. .

Ask, **What other trigonometric identity have you seen involving the sine and the cosine?** As a hint, tell students to think of an identity that expresses the sine of an angle in terms of the cosine of a related angle. As a further hint, draw a right triangle and ask how the sine of one base angle might be expressed as the cosine of another angle, and what the relationship is between the two angles. Use the fact that the angles are complementary to review the equation sin θ = cos (90° − θ).

Then ask, **Does this relationship hold true for all angles?** You may simply have students verify it for specific values of θ in different quadrants.

One possible proof of this relationship uses a quadrant-by-quadrant analysis. A more intuitive approach uses the idea that the graph of the function y = cos θ can be obtained by reflecting the graph of the function y = sin θ about the line θ = 45°.

Key Questions

Does the relationship $\cos^2 \theta + \sin^2 \theta = 1$ require you to use points on the unit circle?

What other trigonometric identity have you seen involving the sine and the cosine?

Does this relationship hold true for all angles?

Supplemental Activity

A Shift in Sine (*extension*) has students examine in more depth how the graph of the cosine function can be produced by shifting the graph of the sine function.

More Pythagorean Trigonometry (*extension*) asks students to clearly explain the Pythagorean identity just developed and to develop similar identities for the remaining four trigonometric functions.

Coordinate Tangents

Intent

This activity continues the process of extending the trigonometric functions beyond their right-triangle definitions.

Mathematics

In this activity, students extend the definition of the tangent function to all angles and construct its graph.

Progression

Students work on the activity individually. The discussion highlights the difference between the period of the tangent function and that of the sine and cosine functions, and emphasizes that the tangent is not defined for all values.

Approximate Time

25 to 40 minutes for activity (at home or in class)

15 to 20 minutes for discussion

Classroom Organization

Individuals, followed by whole-class discussion

Doing the Activity

In the first question of *Coordinate Tangents*, students see that the right-triangle-based definition of the tangent as $\frac{y}{x}$ works for angles in all four quadrants.

Question 2 leads to the identity $\tan\theta = \frac{\sin\theta}{\cos\theta}$. Students test this identity in the third question, and then sketch a graph of $z = \tan\theta$.

Discussing and Debriefing the Activity

Let a volunteer present Question 1, but be sure to determine how well students were able to handle this question. Some may have considered it easy after the work with sine and cosine, but others may have had trouble. Use a first-quadrant example to illustrate that the ratio $\frac{y}{x}$ comes from the right-triangle definition and that it makes sense to define $\tan\theta$ in general using this ratio.

Ask, **Are there any difficulties that could arise from using this ratio to define the tangent function?** If a further hint is needed, ask what happens if $x = 0$. Bring out that the ratio is undefined in that case, and tell students that the tangent function is thus undefined for certain angles.

Have the class determine for which angles the tangent function is undefined. They should see that it is undefined for 90° and 270° in the "first cycle." They might see, more generally, that it is undefined for any odd multiple of 90°. Bring out that for right triangles, the ratio $\frac{\text{opposite}}{\text{adjacent}}$ gets larger and larger as the base angle gets closer to 90°, so it makes sense that there would be a problem at 90°.

On Question 2, the goal is to develop the identity $\tan \theta = \frac{\sin \theta}{\cos \theta}$. Students might get this by writing x and y in terms of r and θ, so that the ratio $\frac{y}{x}$ becomes $\frac{r \sin \theta}{r \cos \theta}$, which simplifies to $\frac{\sin \theta}{\cos \theta}$.

The examples in Question 3 should be fairly direct applications of the definition. If some students were unable to develop the general definition for the tangent function, you may want to have the class work on Question 3 now. The intent is for students to use right-triangle diagrams and reference angles to find the rectangular coordinates of appropriate points, and to apply the general definition. You can have them verify that their calculators give the same answers.

Question 4

On Question 4, the main goal is to have students see the general pattern of the graph. Take this opportunity to review that the tangent function is undefined for certain angles. Also focus on the sign of the tangent function in the different quadrants. You might make a diagram like this to go with similar diagrams for sine and cosine:

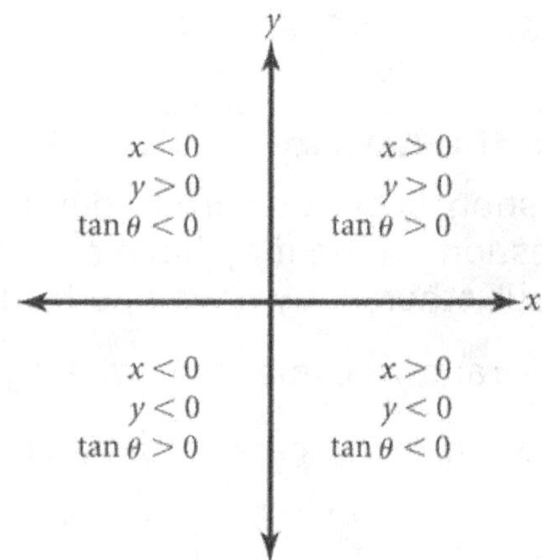

The graph itself should look something like this:

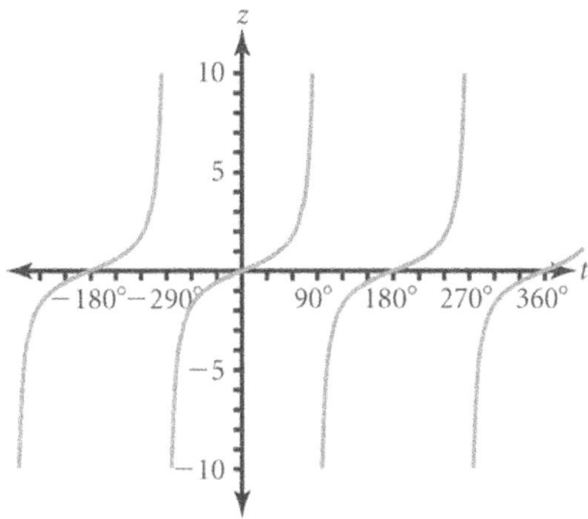

Ask, **What is the period of the tangent function?** Students may assume that the period is 360°, as with sine and cosine. Bring out that the tangent function actually has a period of only 180° because of the way the signs work out.

Key Questions

Are there any difficulties that could arise from using this ratio to define the tangent function?

What is the period of the tangent function?

Positions on the Ferris Wheel

Intent

In this activity, students develop another trigonometric identity.

Mathematics

Students use the Ferris wheel analogy again in this activity to develop the identity $\sin \theta = \sin (180° - \theta)$.

Progression

Students work on the activity individually or in groups, then share their results as a class.

Approximate Time

20 to 25 minutes for activity (at home or in class)

5 minutes for discussion

Classroom Organization

Small groups or individuals, followed by whole-class discussion

Doing the Activity

You may want to go over the "reading" portion of this as a whole class, and then have students begin work on the specific questions.

The activity first asks students to express the angle of rotation for a Ferris wheel position in the second quadrant in terms of the angle of rotation θ for the position in the first quadrant that is at the same height. Students then use the fact that both positions are at the same height to develop the identity $\sin \theta = \sin (180° - \theta)$.

Discussing and Debriefing the Activity

For Question 1, students need to see that the angle for point *B* is $180° - \theta$. Be sure to get an explanation for this conclusion. For instance, students might see that the angle between the ray through *B* and the negative end of the *x*-axis must be equal to θ, as shown here. (They might explain this using the two right triangles in the diagram.)

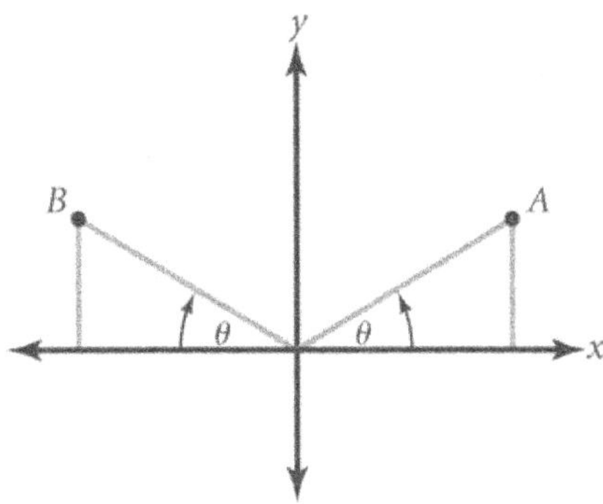

For Question 2, students must set the expressions 50 sin θ and 50 sin (180° - θ) equal to each other. Presumably, they will then divide by 50 to get the identity

$$\sin \theta = \sin (180° - \theta)$$

You may want to post this result for reference in later units.

More Positions on the Ferris Wheel

Intent

This activity continues the work of *Positions on the Ferris Wheel* in developing trigonometric identities.

Mathematics

In this activity, students develop the identities $\cos(-\theta) = \cos\theta$ and $\sin(-\theta) = -\sin\theta$.

Progression

Students work on the activity individually and then share their results as a class.

Approximate Time

25 to 30 minutes for activity (at home or in class)

10 to 15 minutes for discussion

Classroom Organization

Individuals, followed by whole-class discussion

Doing the Activity

Part I of this activity has students explain why the Ferris wheel positions for angles of rotation of θ and $-\theta$ have the same x-coordinate, first for a specific example and then for the general case. They use this information, in turn, to explain why $\cos(-\theta) = \cos\theta$.

In Part II, students are given the identity $\sin(-\theta) = -\sin\theta$. They are asked to check the identity using specific values, and then to explain it using both the Ferris wheel situation and a more general coordinate system diagram.

Discussing and Debriefing the Activity

You can have different volunteers present their ideas on each of the questions in the activity.

Part I: Clockwise and Counterclockwise

Part I is similar to the previous activity, *Positions on the Ferris Wheel*. For Question 1a, students might use a diagram like the following, with C and D

representing the positions of the two riders. They can show that the two right triangles are congruent (or use a more intuitive argument) to explain why *C* and *D* have the same *x*-coordinate.

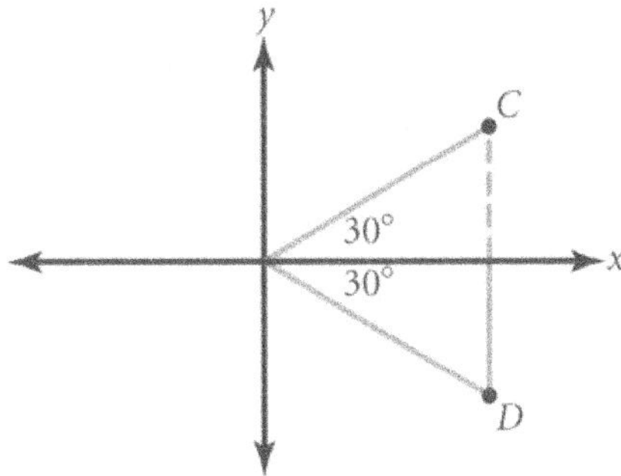

To answer Question 1b, they should reason that the formula gives 50 cos 30° and 50 cos (−30°) as the two *x*-coordinates. Because these *x*-coordinates are equal, cos 30° and cos (−30°) must be equal.

This diagram simply generalizes the previous one:

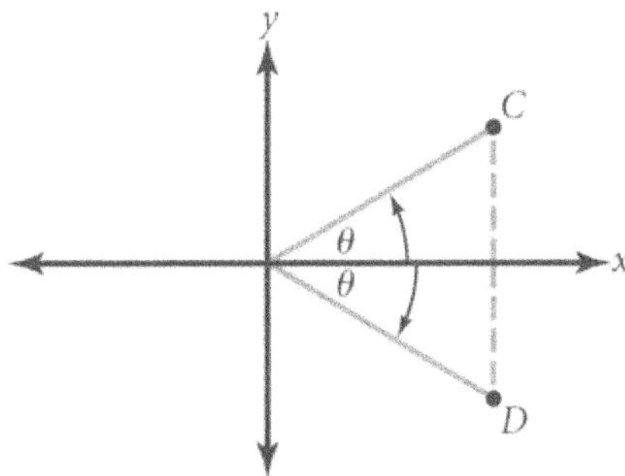

The explanation for Question 2a is essentially the same as that for Question 1a. For Question 2b, students will probably have little difficulty moving from their work in deriving the equation cos (−30°) = cos 30° to the more general identity

$$\cos(-\theta) = \cos\theta$$

Ask, **How is this equation connected to the graph of the cosine function?** Help students see that the equation says that the graph of the function $y = \cos\theta$ is symmetric about the *y*-axis.

Part II: From Identity to the Ferris Wheel

For Part II, begin by having students illustrate their work from Question 3, explaining the identity $\sin(-\theta) = -\sin\theta$. Be sure to include an example involving a negative angle, because the issue of signs is important here. Specifically, they should see that if θ itself is negative, then $-\theta$ is positive.

The main task of Part II is finding a Ferris wheel situation that expresses the given identity. Let volunteers offer ideas. For example, a student might use the diagram from Part I and point out that a rider going counterclockwise is just as high above the center of the Ferris wheel as a rider going clockwise is below the center.

Post all of the identities discussed in this activity.

Key Question

How is this equation connected to the graph of the cosine function?

A Trigonometric Reflection

Intent

In this activity, students summarize ideas about trigonometry.

Mathematics

A Trigonometric Reflection asks students to compile a summary, complete with diagrams and explanations, for the trigonometric concepts from this unit. Those ideas include extension of the trigonometric functions beyond first-quadrant angles, graphs and periodicity of the functions, identities, and polar coordinates.

Progression

This activity will be included in the students' portfolios for this unit.

Approximate Time

30 minutes for activity (at home or in class)

20 minutes for discussion

Classroom Organization

Individuals, followed by whole-class discussion

Doing the Activity

Tell students that this summary will be included in their portfolios for this unit.

Discussing and Debriefing the Activity

You may want to focus on the process students used in extending the trigonometric functions from the right-triangle definitions to the complete functions defined in this unit.

High Dive Portfolio

Intent

In this activity, students compile their unit portfolio.

Mathematics

In addition to the unit portfolio, the discussion here includes opportunities for students to summarize the unit and to speculate on what else might need to be considered in order to make the unit problem more realistic.

Progression

For the portfolio, students write a cover letter in which they summarize the main mathematical concepts studied in this unit, and they select samples of their work that were instrumental in developing those ideas. They engage in class discussion speculating on why the situation presented in the unit problem is actually more complex than what they have worked with so far. The class also summarizes what has been learned in this unit.

Approximate Time

30 minutes for activity (at home or in class)

30 minutes for speculation on the more complex version of the unit problem

10 to 20 minutes for unit reflection

Classroom Organization

Individuals, followed by whole-class discussion

Doing the Activity

Students work independently writing their cover letter and selecting portfolio activities that reflect the mathematical ideas learned in the unit.

Discussing and Debriefing the Activity

Ferris Wheel Speculation

To close the unit, you might have students speculate on the more complex version of the unit problem.

If this hasn't come up before, explain that when the diver is released from the platform, he leaves with the same speed and direction of motion that the platform itself has at that instant. For example, if he is released at the 9 o'clock position, at that moment the platform is moving downward, and the diver will actually start out with some non-zero speed in the downward direction, and will reach the ground (or the water level) sooner than if he had been dropped from a stationary platform. If he is released at the 12 o'clock position, at that moment the platform is moving to the left, and the diver will start out with some non-zero speed in that direction, and will end up to the left of the center of the Ferris wheel.

You can tell students that taking this motion into account makes the problem considerably more complex, and that solving this version of the problem is the focus of the Year 4 unit *The Diver Returns.*

Next, let volunteers share their portfolio cover letters as a way to start a discussion to summarize the unit.

Blackline Masters

The Circus Act

The Ferris Wheel

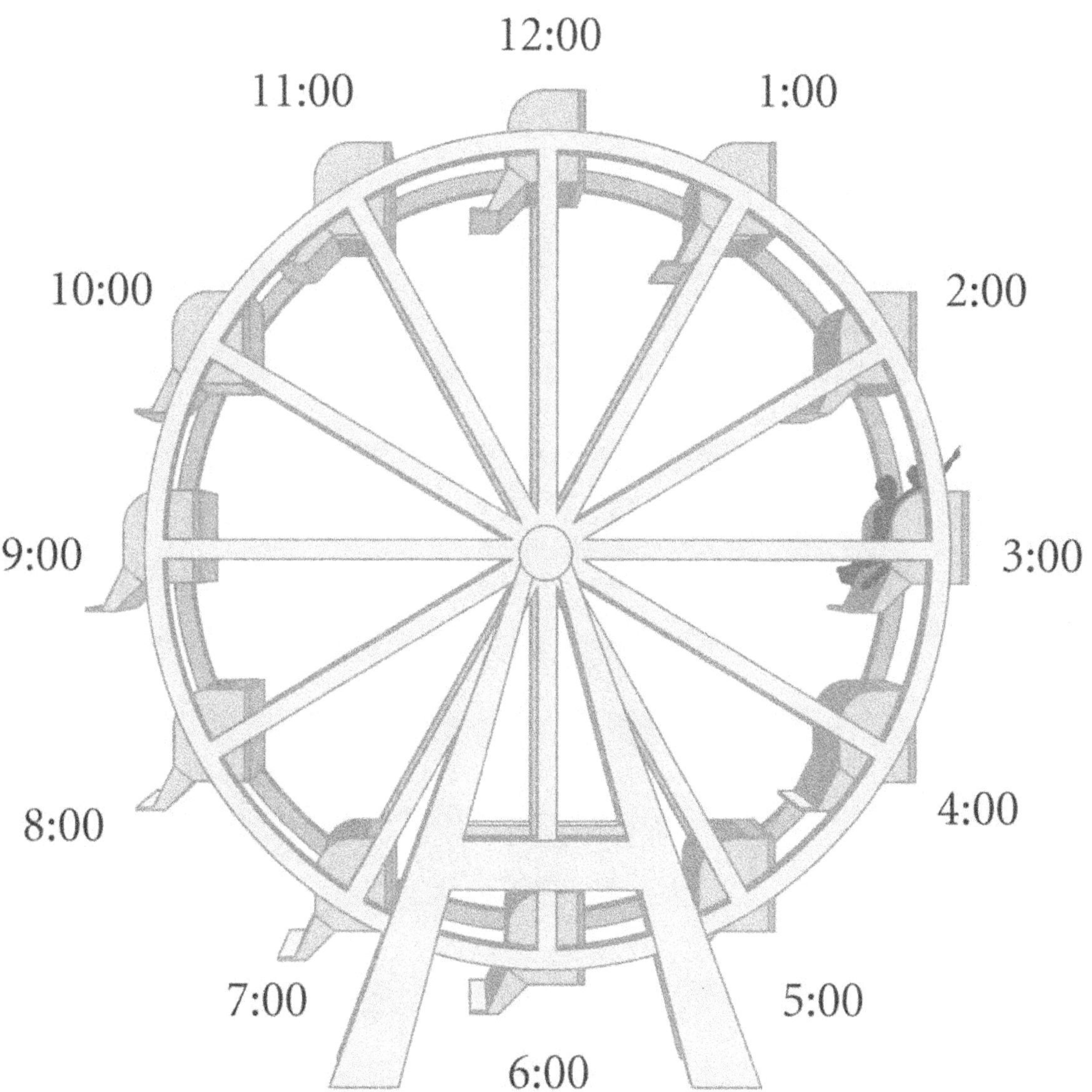

Graphing the Ferris Wheel

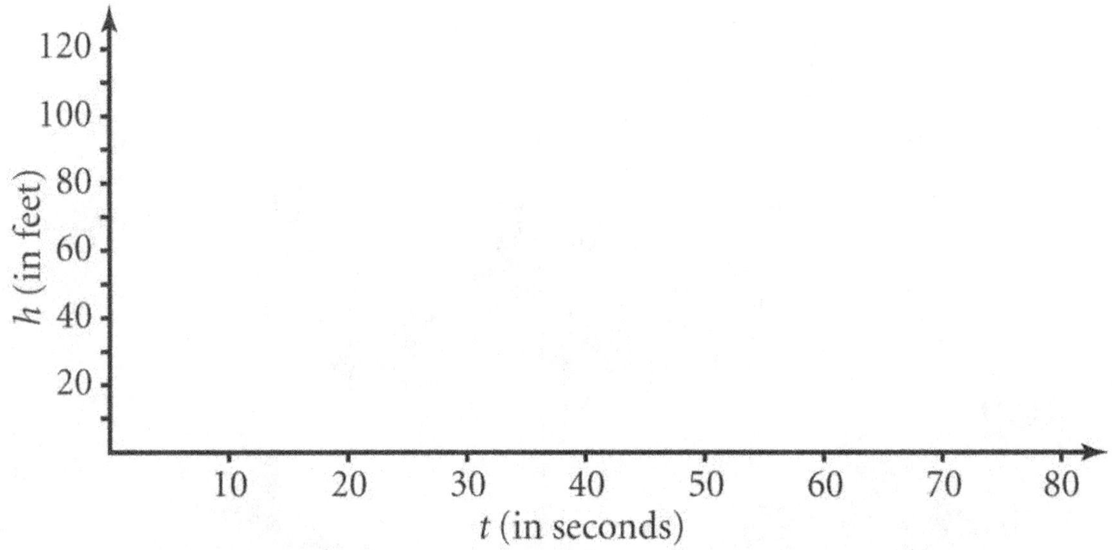

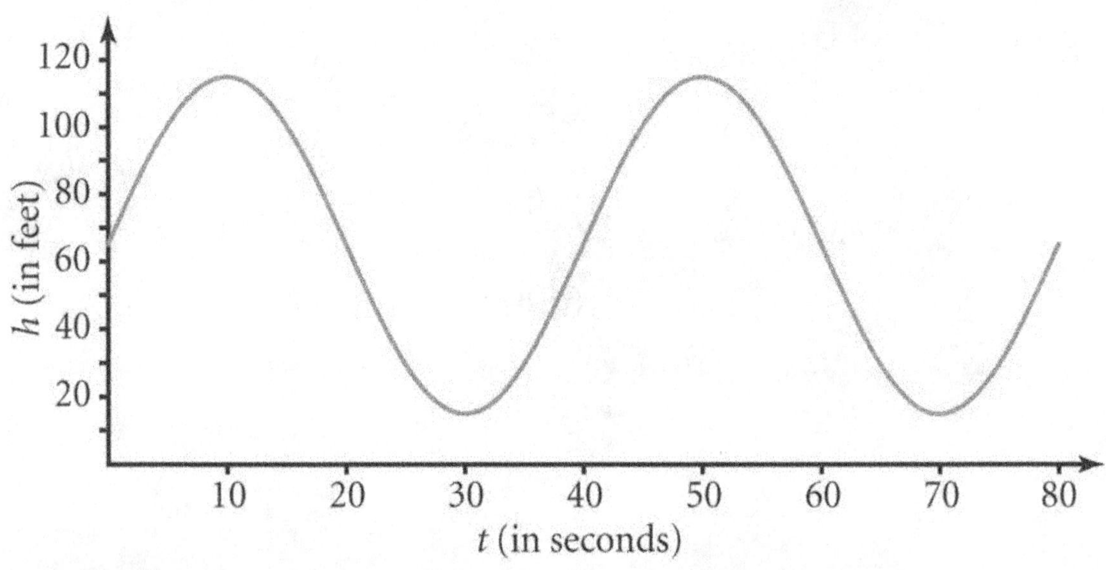

The "Plain" Sine Graph

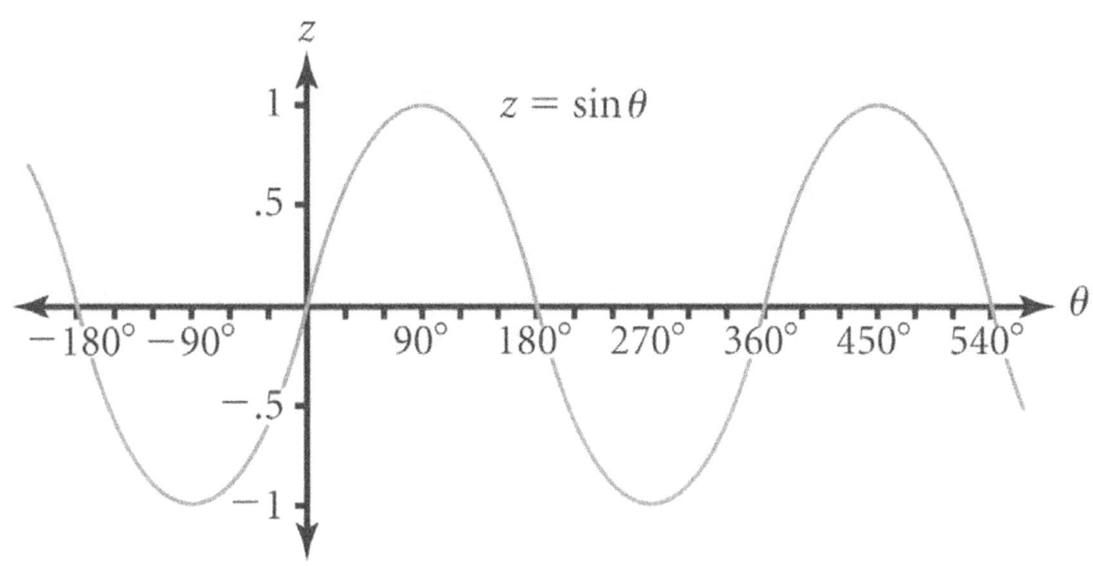

Distance with Changing Speed

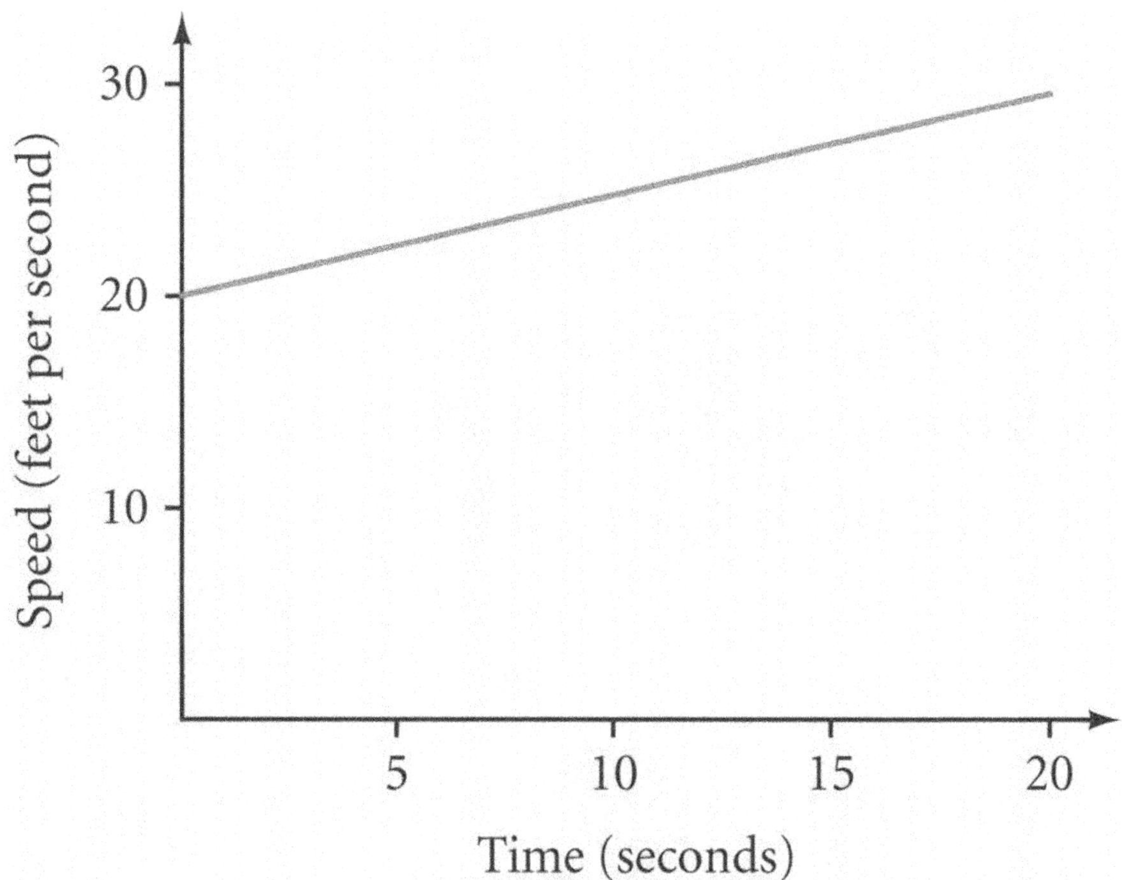

Where Does He Land?

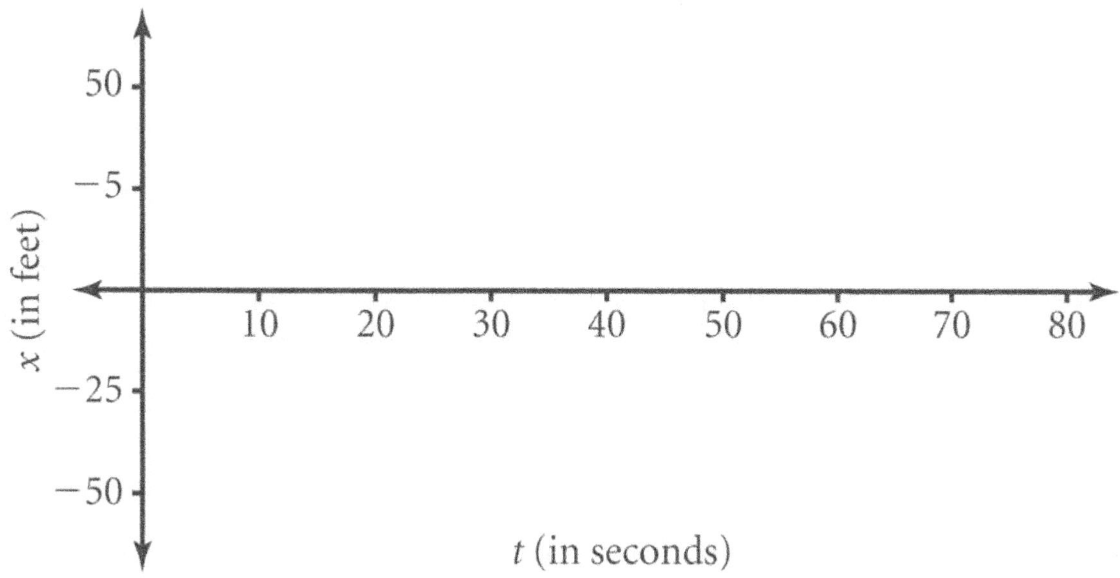

¼-Inch Graph Paper

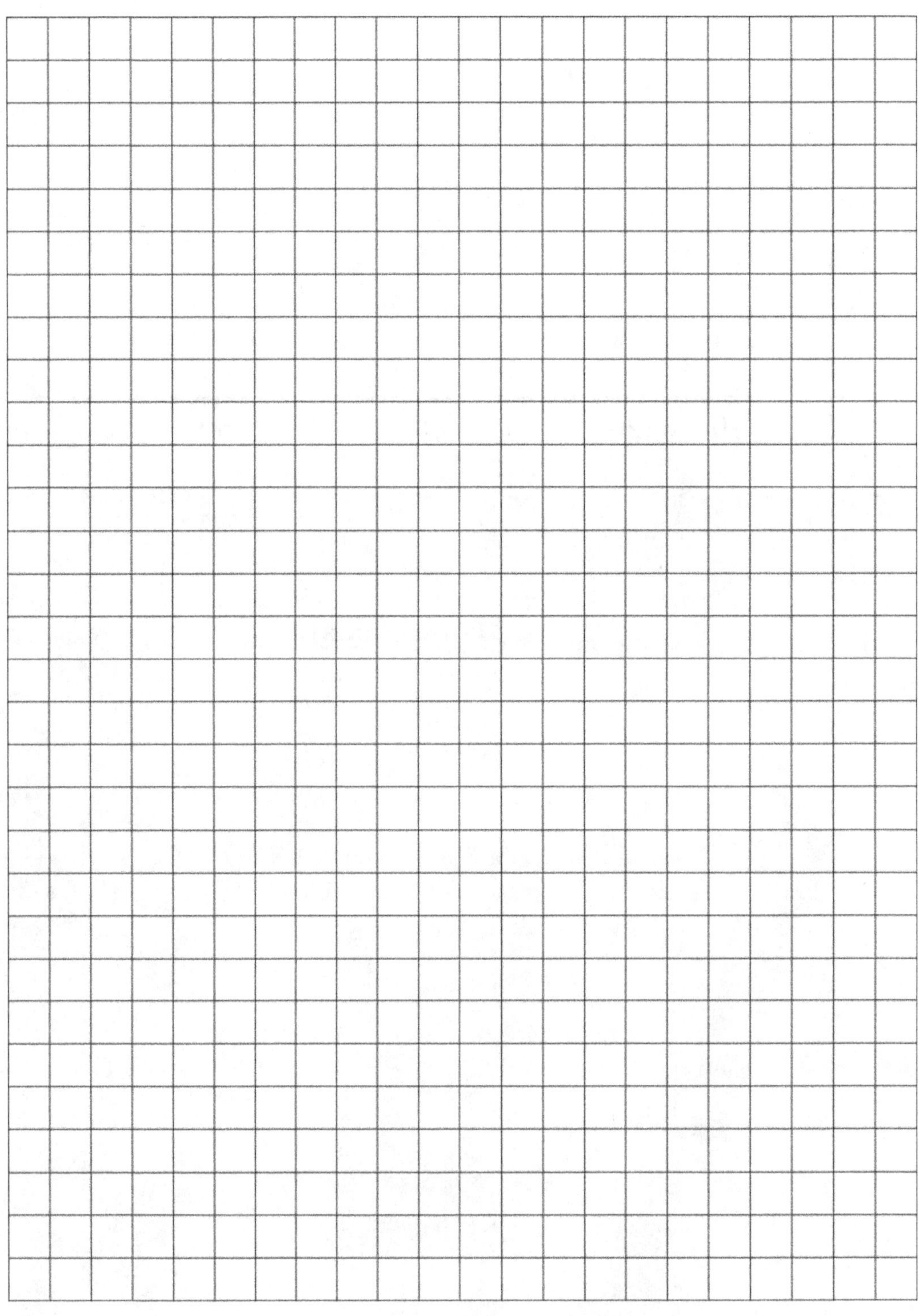

1-Centimeter Graph Paper

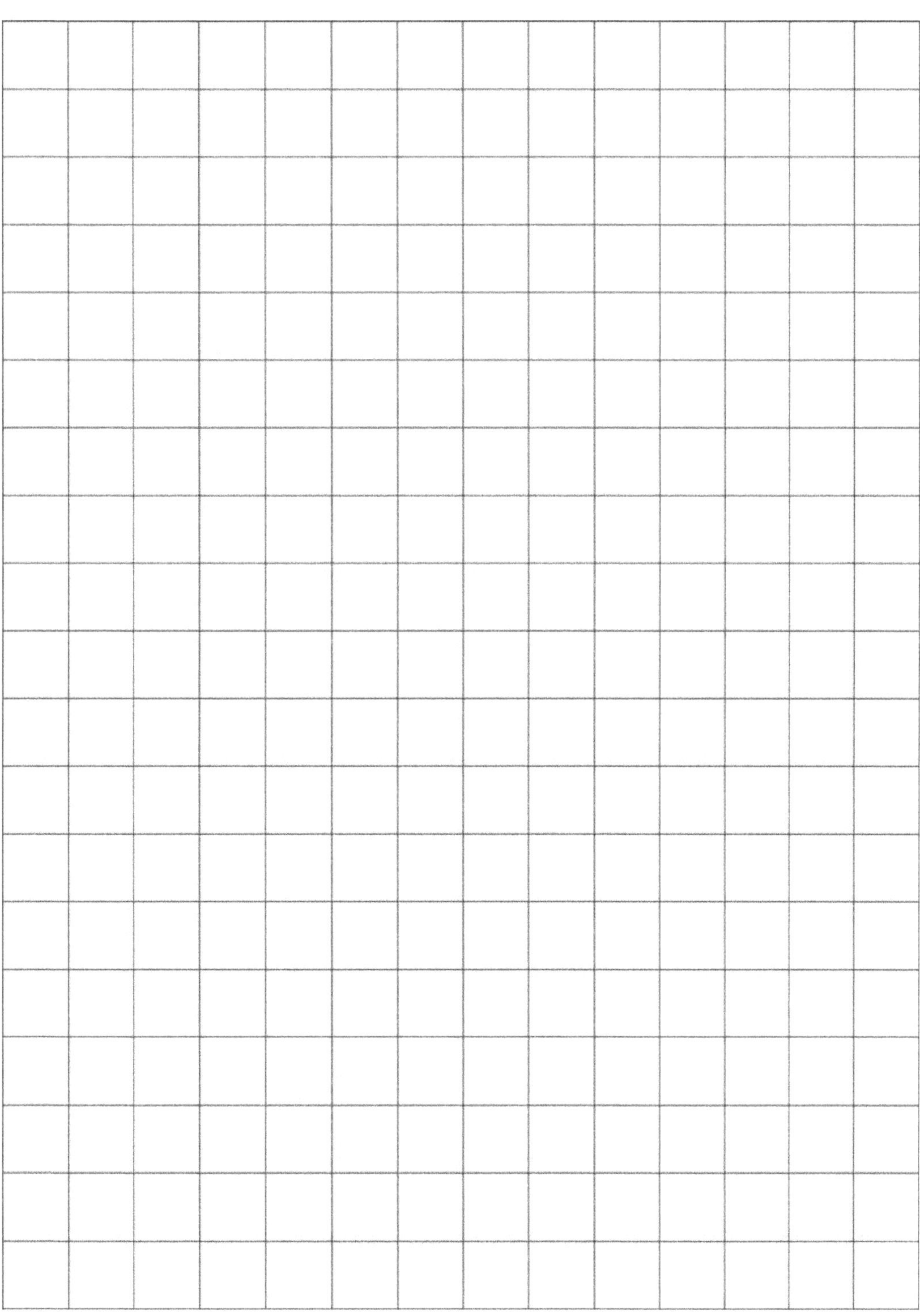

1-Inch Graph Paper

Assessments

In-Class Assessment

Walter the whale is a mathematical sort of creature. He swims with a periodic rise and fall, patterning his swimming path after a sine curve.

On a particular day, Walter is swimming with amplitude of 15 feet. More specifically, when he is at his highest point, his back rises 5 feet out of the water. Then he dives down to a point where his back is 25 feet below the surface level.

Of course, he then comes back up again, and then goes down again, and so on. It takes about 20 seconds from the time Walter hits his high point until he reaches his lowest point.

You come out on the deck of a ship and look out to sea exactly in the direction where Walter is swimming. What is the probability that you will see him the instant you look?

Take-Home Assessment

1. Here is a stone arch in the shape of a semicircle. A pebble located at polar coordinates (80, 125°) works loose and falls. How long will it take the pebble to hit the ground?

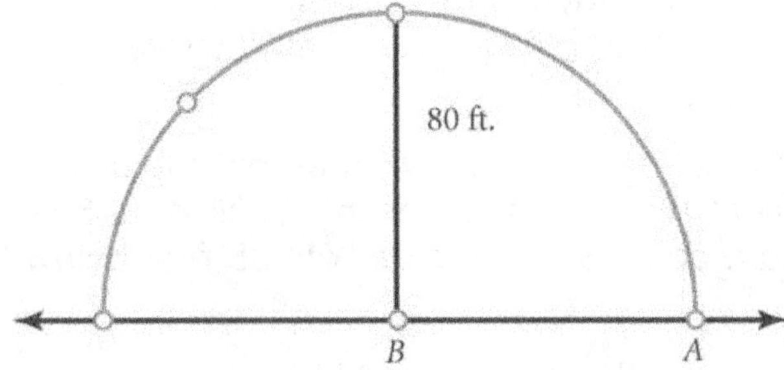

2. In this unit, you've seen that the sine, cosine, and tangent functions can be extended to arbitrary angles using the coordinate system and a diagram like the one below. Explain how to use the diagram to define the *secant* function for all angles.

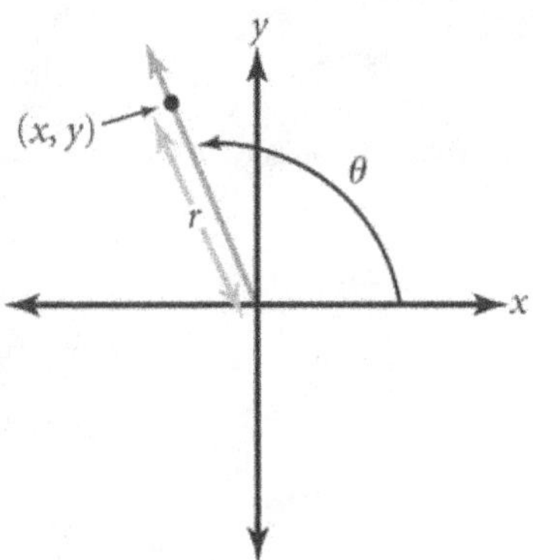

Reminder: In a right triangle like the one shown here, the secant function is defined by the equation $\sec \theta = \frac{AB}{AC}$.

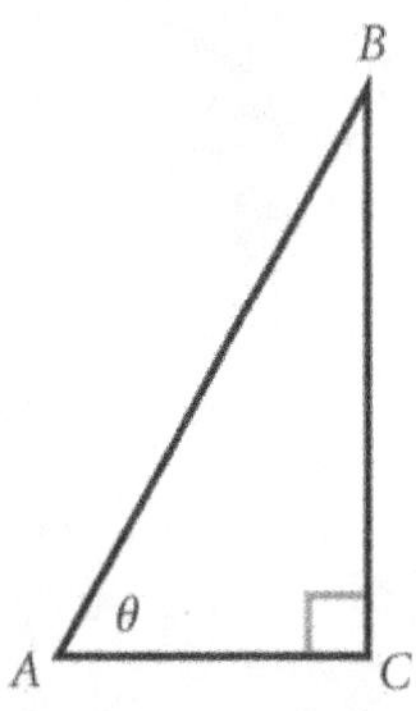

IMP Year 3
First Semester Assessment

I. Once Upon a Time . . .

Imagine that you are Madie or Clyde. You've grown old and are telling your grandchildren the story of the orchard hideout. You've described the arrangement of trees in the original orchard (with a radius of 50 units) and told them the basic facts that you knew at the start.

- The circumference of the newly planted trees
- The fixed amount by which the cross-sectional area of the trees grew each year
- The distance between the centers of adjacent trees

Of course, the grandchildren have heard the story before, and they remember that it took about 11 years, 9 months for the center to become a true orchard hideout. What you want to do is impress them with how well you and your partner analyzed the problem back then.

Write a description of how the analysis worked. Don't get bogged down in the specific numbers, because you don't have pencil and paper handy, and the youngsters are more interested in the big ideas anyway!

II. Road Building

The highway department is planning a road that will go through the town of Coldwater. The town of Hot Springs is 13 miles due south of Coldwater, and the town of Warm Rock is 18 miles due east of Hot Springs.

1. Sketch a diagram showing the relationship between the three towns. (Treat each town as a single point.)

The mayors of Hot Springs and Warm Rock both want this new road through Coldwater to go straight through their towns as well. Unfortunately, the highway department can afford to build only one road.

The road must go through Coldwater and must be straight, so a compromise route is needed. The mayors of Hot Springs and Warm Rock agree to support the project if this condition is met:
The distance from Hot Springs to the new road must be the same as the distance from Warm Rock to this road.

They also insist that the road should not be parallel to the route from Hot Springs to Warm Rock.

2. a. Add a dotted line to your sketch from Question 1 to show where the new road must go, and explain your reasoning.

b. Find the distance from Hot Springs to the road, to the nearest tenth of a mile, and explain your work.

III. Equation Time

Solve this system of equations, and explain your work.

$$7r + 6s = 6$$
$$5r - 4s = 25$$

IV. The Third Dimension

This system of linear constraints in three variables defines a feasible region.

I	$2x + y + z \leq 20$
II	$3x + 4z \leq 13$
III	$y + 2z \leq 9$
IV	$x \leq 9$
V	$y \leq 6$
VI	$x \geq 0$
VII	$y \geq 0$
VIII	$z \geq 0$

Give *a general outline* of how to find the point in this feasible region where the function $2x + y + z$ has its maximum, and explain the geometric reasoning behind your method.

V. Solving with Matrices

Consider this system of linear equations.

$$3a + 2b - c + d = 1$$
$$2a - b + 4c + 2d = -2$$
$$-4a + 3c - 3d = -6$$
$$a + b + c + d = 3$$

1. Write a matrix equation that is equivalent to this system.

2. Solve the system using matrices on a graphing calculator, and show your solution.

3. Discuss the relationship between the matrices and the equations and the properties of matrices that allow you to use them to solve systems of linear equations.

IMP Year 3
Second Semester Assessment

I. Spilt Milk

You've automated your dairy farm so that all the cows are milked by milking machines, and the milk all flows into one giant cone-shaped container. At the start of milking time, the container is empty, and as the milk flows in, the level in the container rises. Milking starts at 5:00 a.m. and continues through the day. (The cows are not all milked at the same time.)

After studying your cows and using some geometry, you've figured out that at t minutes after 5:00 a.m., the milk in the container will have risen to a level of $\sqrt[3]{2000t}$ centimeters.

1. During the hour from 7:00 a.m. until 8:00 a.m., what is the *average* rise per minute in the height of the milk? (Give your answer to the nearest 0.001 cm/min.)

2. At what rate is the milk level rising at 8:00 a.m.? (Again, give your answer to the nearest 0.001 cm/min.)

3. At what time of day will the milk level reach 100 centimeters?

II. Darts

Consider a square dartboard with a circle inscribed in the square, as shown here.

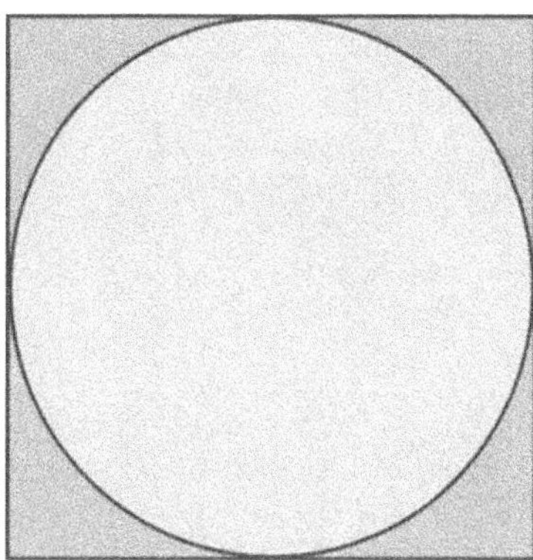

Suppose that according to the rules, if your dart lands inside the circle, you win, and if the dart lands outside the circle, you lose. Assume that you always hit the dartboard and that each point of the square is equally likely to be hit.

1. If you throw one dart, what is your probability of winning? Explain your answer, giving the probability to the nearest hundredth.

2. Suppose you throw seven darts. What is the probability that you will win at least four times? Explain what method you use to find the answer and why the method works. Again, give the probability to the nearest hundredth.

III. Ferris Wheel Fence, Revisited

It's time to look back at the problem of the fence around the amusement park, from *High Dive*.

As you may recall, Al and Betty are riding on a Ferris wheel. This Ferris wheel has a radius of 30 feet, and its center is 35 feet above ground level. There is a 25-foot-high fence around the amusement park, but once you get above the fence, there is a wonderful view.

What percentage of the time are Al and Betty above the level of the fence?

IV. Opposite Angles

You have learned these formulas involving trigonometric functions:

$$\cos(-\theta) = \cos\theta$$
$$\sin(-\theta) = -\sin\theta$$

Explain each of these formulas in several ways:

- In terms of the Ferris wheel
- In terms of the graphs of the sine and cosine functions
- Using numerical examples

You can use these graphs of sine and cosine in your explanation:

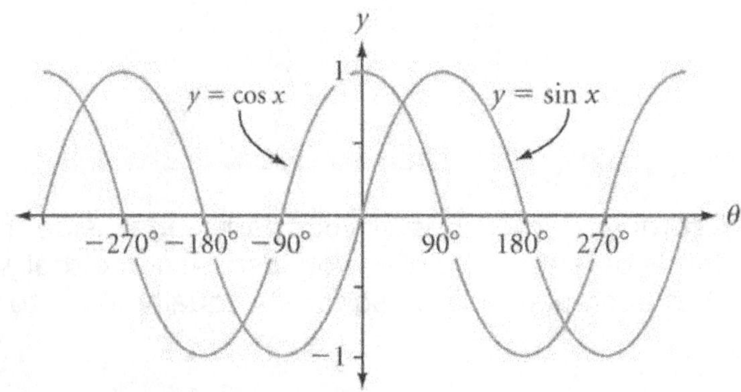

You can also use this diagram to represent a Ferris wheel:

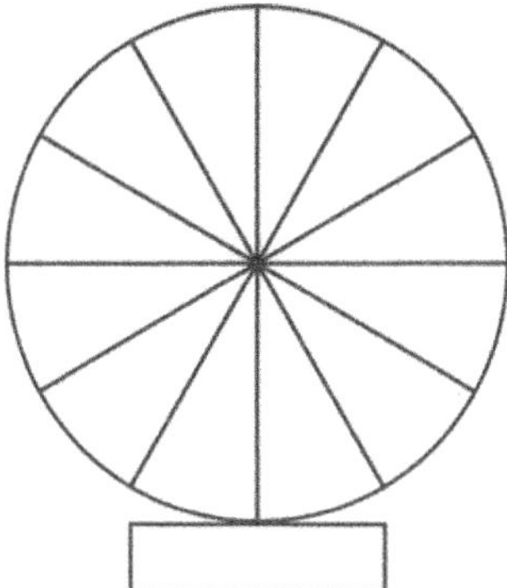

High Dive Calculator Guide for the TI-83/84 Family of Calculators

High Dive makes use of what students already know about the graphing calculator and introduces several new techniques as well. The unit problem itself leads to an equation that is far too complex to solve by algebraic manipulations and that instead is solved graphically with the calculator. Even a graphical solution without the calculator would be awkward. Students should gain a tremendous sense of satisfaction at having solved an impossible-looking equation, as well as an appreciation for the power of the graphing calculator.

Throughout the unit, students can use the calculator to check graphically what they discover with algebra and trigonometry. This unit also provides an appropriate context in which to reveal the CALC menu, which is full of helpful features. Near the end of the unit, as students pull together the many elements of the unit problem, they will learn how to break lengthy functions into smaller pieces for entry into the calculator.

POW 1: The Tower of Hanoi: The recursive pattern for *POW 1: The Tower of Hanoi*, $a^{n+1} = 2a^n + 1$, can easily be illustrated on the calculator. Enter the number of moves needed for a single disc by pressing [1] [ENTER]. Find the number of discs for the second move by pressing [×] [2] [+] [1] [ENTER]. Press the [ENTER] key again to calculate the number of moves for a tower of three discs. Each time the [ENTER] key is pressed, the total for the next larger tower will be displayed.

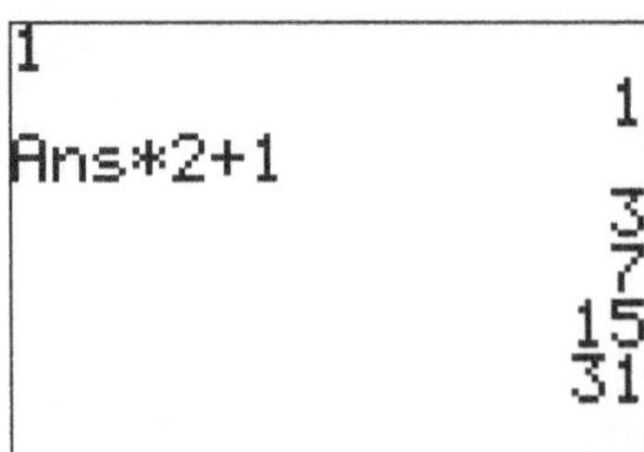

The Ferris Wheel: When reviewing *The Ferris Wheel,* you might remind students to select [MODE] and to check that their calculators are in degree mode. The calculator defaults to radian mode whenever the calculator is reset. This can happen when the batteries are low or are changed. Students should develop the habit of checking the mode if they get results that seem odd, or whenever they use trigonometric functions.

Because students will be using trigonometric functions for many of the activities in this unit, encourage them to explore how the functions work on their personal calculators. For example, on the graphing calculator, you find sin 80° by pressing [SIN] [8] [0] [ENTER], in that order. On most scientific calculators, the order is reversed; that is, the angle must be entered before the sine function is selected.

You might also remind students to be careful with placement of parentheses. As shown here, the graphing calculator automatically opens a set of parentheses when a trigonometric function is selected. Unless those parentheses are closed, the calculator assumes closure at the end of the line. Compare the results to the two expressions entered here.

```
15sin(30+15+5
     11.49066665
15sin(30)+15+5
            27.5
```

As the Ferris Wheel Turns: Although an approximation of 3.14 for π may be close enough for this activity, remind students that π is an irrational number and can be obtained more accurately by using [2ND] [π] on the calculator. The activity *Falling Bridges,* in the Year 2 unit *Do Bees Build It Best?,* illustrated that small round-off errors can produce dramatic differences in the final result under some circumstances. Encourage students to develop calculator habits that minimize those differences.

In both *As the Ferris Wheel Turns* and the subsequent activity, *At Certain Points in Time,* placement of parentheses may again become an issue. Students need to remember to close parentheses that the calculator opens automatically. Thus, in Question 4b of *As the Ferris Wheel Turns,* the parentheses cannot be omitted from **65+50sin(9*6)**. Compare the results of the calculations shown in this display.

```
65+50sin(9*6)
     105.4508497
65+50*sin(9)*6
     111.9303395
```

A Clear View: Discussion of *A Clear View* offers an excellent opportunity to remind students of the ease with which previous answers can be recalled for use in the calculator, rather than working with rounded approximations. It will make only a slight difference in this activity, but these habits will be important in other instances, particularly for students pursuing careers in science or engineering.

```
cos⁻¹(7/15)
     62.18186072
360-2*Ans
     235.6362786
Ans/360
     .6545452182
```

Graphing the Ferris Wheel: After bringing out that individual points for Question 1 of *Graphing the Ferris Wheel* can best be found with the equation $h = 65 + 50 \sin 9t$, ask students how they went about constructing their graphs. Most probably created an In-Out table in preparation for graphing. Ask how they might have used their calculators to create this table more quickly.

Instructions for using the TABLE feature is provided in the Calculator Note "Creating Tables on the Calculator." In Step 4, students should discover that it is necessary to set **TblStart** to zero. A value of 5 for **ΔTbl** will yield a sufficient number of values at convenient intervals for graphing by hand.

After viewing students' graphs for Question 1, have them graph this function on their calculators. (But we recommend that calculator graphs for the situations in Question 2 not be generated until the discussion of *Ferris Wheel Graph Variations*.) Students should see that the graph matches the one they created by hand. The procedure for graphing this on the calculator should be familiar to most Year 3 students, but instructions follow for those who need review or may not have experience with the graphing calculator. See the Calculator Note "Graphing Functions."

The set of values **Xmin=–5, Xmax=80, Ymin=–5,** and **Ymax=120** will yield a nice window for display of this graph for Question 1, but resist the urge to simply give these values to students. They need the experience that comes from thinking about what the two variables represent and deciding on an appropriate range of values. If necessary, remind them that they made this same decision when they decided how to scale the graphs they drew by hand.

While discussing the periodicity of the graph in Question 1, you might have a student use the graph to illustrate what a period of 40 seconds means. This can be seen approximately on the calculator simply by pressing [TRACE] and using the left and right arrows to move the cursor along the graph. You may not be able to find *x*-values that are exactly 40 units apart in this way, due to the size of the increment assigned to each pixel on the screen, but a good approximation should suffice. Students can also use the number keys to enter the exact *x*-coordinate in which they are interested, after pressing [TRACE]. The Calculator Note "Tracing Tips" presents instructions for obtaining values for the window variables that yield a friendlier window for tracing.

Ferris Wheel Graph Variations: The calculator provides an ideal way to compare the graphs of the functions students develop in *Ferris Wheel Graph Variations* with the graph of the original function. Have students check their work on this activity by graphing each of the functions on the calculator, as described in the Calculator Note "Graphing Functions." Tell them to enter the original function from *Graphing the Ferris Wheel* as $\mathbf{Y_1}$ in the [Y=] screen. They should then enter the equation as modified as $\mathbf{Y_2}$.

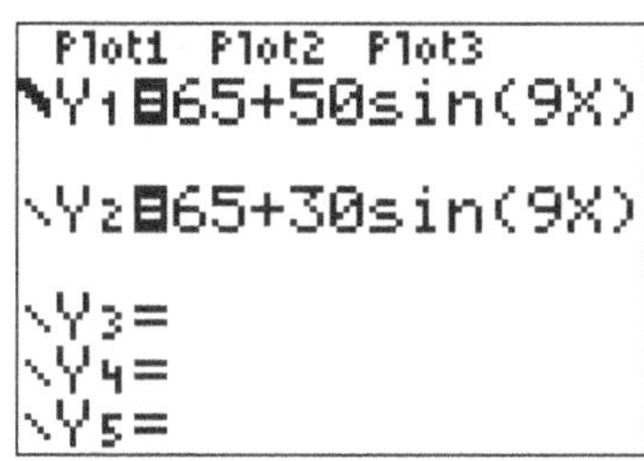

If you like, you can select a heavier line for the graph of the original function to more easily distinguish the two graphs. Do this while still at the Y= screen by using the left arrow to move the cursor to the "\" symbol, which is to the left of **Y1**. Press ENTER to cycle through the various line styles. Pressing ENTER only once will select the heavier "****" symbol.

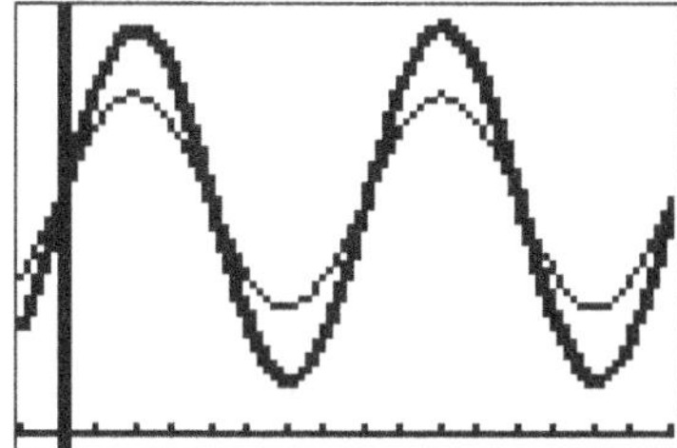

You may also turn off an individual function without erasing it by moving the cursor to the equal sign and pressing ENTER to remove the highlighting. The deselected function will still be evaluated, but it will not be graphed.

The "Plain" Sine Graph: When working on this activity, students will find it helpful to use the TABLE feature of their calculators to obtain the data set to graph. See the Calculator Note "Creating Tables on the Calculator."

Sand Castles: Graphing the function from this activity will be useful for the discussion. Begin by having students graph $w(t) = 20 \sin (29t)$ on their calculators, with a viewing window using scales similar to those they chose when graphing the function by hand. See the Calculator Note "Graphing Functions." Point out again that the TABLE feature of the calculator could have been useful for generating the data set for their graphs.

For Question 2, after (and only after) students are able to explain how they can see from the function that the minimum and maximum values must be -20 and 20, let them verify that with the graphing calculator CALC features. Note that we have avoided helping students discover these features until now, because we did not want to short-circuit the learning of the mathematics behind them. But now, an awareness of these calculator tools will be useful as students move beyond high school. See the Calculator Note "Solving 'Sand Castles' with the CALC Menu."

Portions of Questions 3 and 4 are also covered in the Calculator Note "Solving 'Sand Castles' with the CALC Menu." Once again, use this material only after a thorough discussion focusing on a solution based on the properties of the sine function.

Question 5 requires students to find the coordinates of the point 1 unit to the left of the minimum vertex of the curve. Trying to obtain sufficient accuracy while tracing the graph of the function can be frustrating. The instructions in the Calculator Note "Tracing Tips" are worth discussing even if you elected not to introduce the CALC features. They provide several tips for tracing to an exact point with minimal fuss.

POW 2: Paving Patterns: You can use the calculator to demonstrate that the closed form of the equation from *POW 2: Paving Patterns* really does, amazingly, yield the correct sequence. The function has been entered here using the techniques described in the Calculator Note "Graphing a Complicated Equation." See also the Calculator Note "Creating Tables on the Calculator" for instructions on using the calculator's TABLE feature.

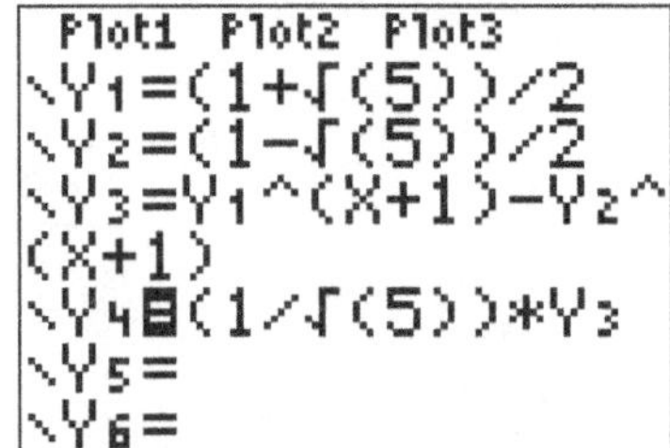

X	Y4
1	1
2	2
3	3
4	5
5	8
6	13
7	21

X=1

More Beach Adventures: Though the CALC features are useful tools, do not let them become a substitute for working with the powerful mathematics in this activity. The goal here is for students to become thoroughly familiar with the properties of the sine function. Learning to use the CALC features must be secondary. Require students to show the mathematics that supports their solution, even if they check it with the help of the CALC features.

The inverse trigonometric functions will be used frequently throughout the remainder of this unit. Remember that $\sin^{-1} 0.4$, for example, is entered as 2ND [SIN^{-1}] . 4 ENTER.

The *Teacher's Guide* suggests that near the end of the discussion of *More Beach Adventures,* you have students see what happens when they ask the calculator to find $\sin^{-1}(2)$. The result is shown below. What an excellent opportunity to reinforce the vocabulary of *domain* and *range*! Selecting **Quit** will return you to a new line on the home screen, ready for a new operation. Selecting **Goto** will allow you to edit the command that caused the error to occur.

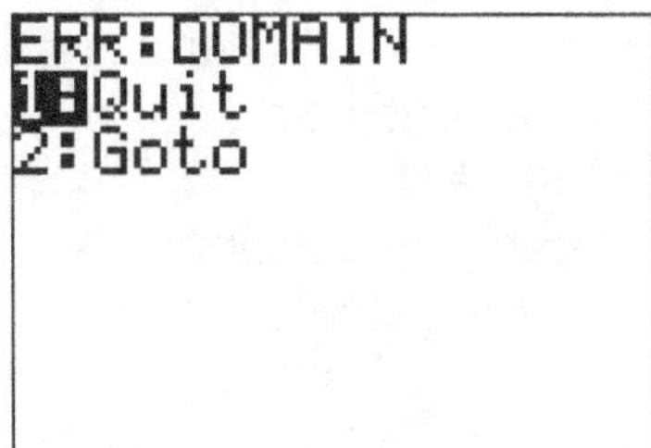

Free Fall: Question 5 of this activity requires students to find an approximate value for $\sqrt{\frac{82}{16}}$. The calculator will automatically open a set of parentheses when the square-root function is selected. If the parentheses are not closed, the calculator will assume closure is at the end of the line.

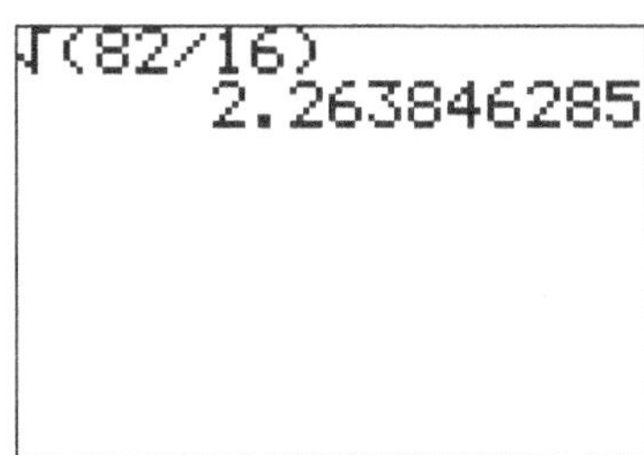

If some students find the answer to this question by using the equation $90 - 16t^2 = 8$, you might ask how they could solve this graphically. If **Y1=90-16X^2** represents the height, they can trace this graph to find the value of **X** that yields a height of 8, or they can find the intersection between this function and **Y2=8**. (See *Question 4: Finding Intersections* in the Calculator Note "Solving "Sand Castles" with the CALC Menu.")

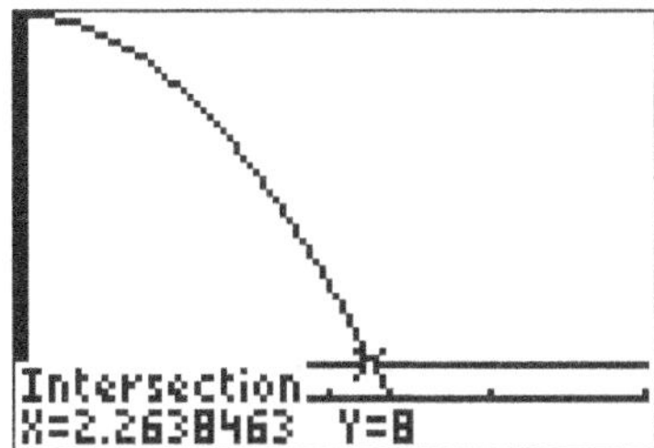

Generalizing the Platform: The *Teacher's Guide* suggests that you have students graph the cosine function on their calculators and compare it with the posted graph of the sine function (see the Calculator Note "Defining the Cosine Function"). The similarity of the two curves can be demonstrated vividly by having students graph both functions at once on the calculator, entering one as **Y1** and the other as **Y2**. You might also move the cursor to the left of **Y1** in the Y= screen and press ENTER to change the "\" symbol to the thicker "****" symbol. This will cause one function to be drawn with a thicker line, making it easier to distinguish between the two overlapping curves.

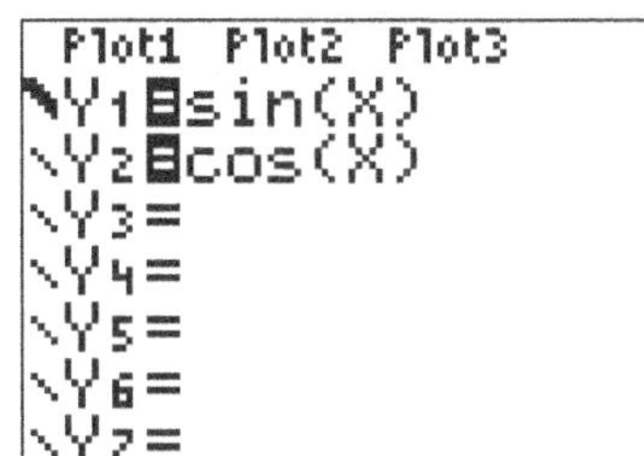

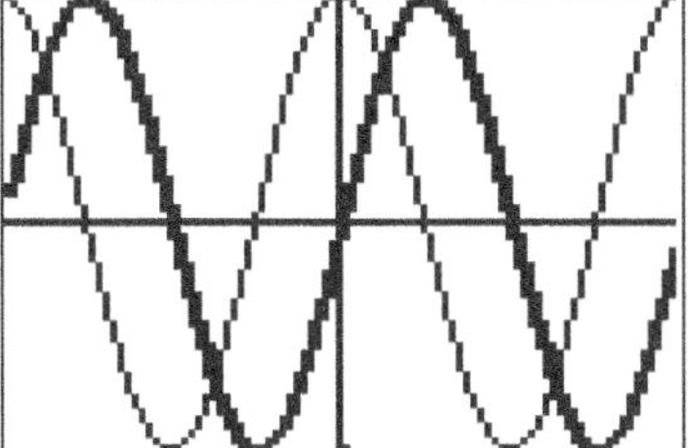

Moving Cart, Turning Ferris Wheel: The solution for this activity will involve some complex calculator work, whether it is solved by guess-and-check, graphically, or with the calculator's SOLVE feature. One of the biggest stumbling blocks will be the correct nesting of parentheses.

Students using guess-and-check will have a difficult time due to the need to enter this complicated equation repeatedly. The Calculator Note "Guess-and-Check Without the Pain" offers some suggestions that involve using multiple commands separated by colons (to avoid the confusion of nested parentheses) and 2ND [ENTRY] (to avoid having to reenter the complex commands for each guess).

The screen demonstrates the results of a guess of $W = 12.28$ for the equation $-240+15\left(W+\sqrt{\frac{57+50\sin 9W}{16}}\right)-50\cos 9W=0$. Notice that the approximate guess does not yield a result of exactly zero, but is essentially correct.

```
12.28→W:(57+50si
n(9W))/16→A:-240
+15(W+√(A))-50co
s(9W)
     -.0623674989
```

Students using a graphical approach will be aided by the techniques described in the Calculator Note "Graphing a Complicated Function," which explains how to break a large function into smaller pieces to reduce confusion. The screens shown here illustrate one way to solve the equation $-240+15\left(W+\sqrt{\frac{57+50\sin 9W}{16}}\right)=50\cos 9W$ by looking for the intersection of the graphs of the left and right halves of the equation, represented by **Y3** and **Y4**. Only these two functions are activated for graphing. Once the two functions are graphed, the intersection can be found by tracing or by using the **intersect** feature on the 2ND [CALC] menu.

```
 Plot1 Plot2 Plot3
\Y1=57+50sin(9X)
\Y2=√(Y1/16)
\Y3=-240+15(X+Y2
)
\Y4=50cos(9X)
\Y5=
```

```
WINDOW
 Xmin=-20
 Xmax=20
 Xscl=1
 Ymin=-30
 Ymax=30
 Yscl=1
 Xres=1
```

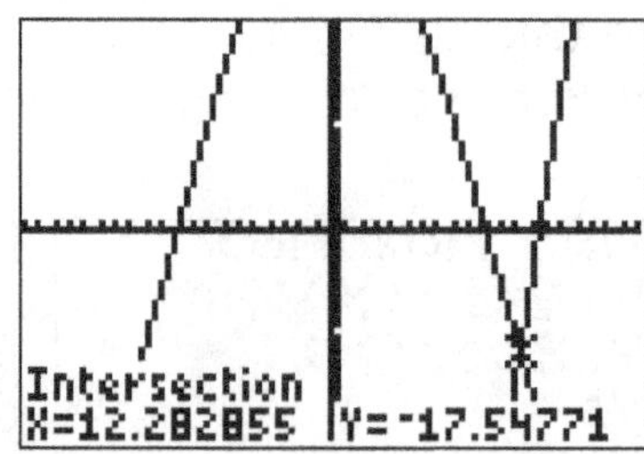

Interpreting the results may be trickier than is initially apparent. Once students have found the coordinates of the intersection, ask them what these two variables represent. Bring out that **X** represents the diver's time on the wheel, in seconds, and is not his x-coordinate, and that **Y** represents the x-coordinate of the diver when he is released from the Ferris wheel. There will likely be some initial confusion.

This activity is likely to require more than a single class period, so you may need to consider how to avoid having the equations from one class revealed to (or destroyed by) students from another class. One solution is to have students save their functions as a graph database, as described in the Calculator Note "Storing and Recalling a Graph Database." After saving the functions as a graph database, have students clear their functions from the Y= screen.

The equation can also be solved using the SOLVE feature. Because the hardest part of this approach is entering the complicated equation, we suggest that you have students do that first, using the instructions from the Calculator Note "Graphing a Complicated Function." Then proceed to the instructions in the Calculator Note "Using the Equation Solver."

```
Y3-Y4=0
▪X=12.282854594…
 bound={0,20}
▪left-rt=0
```

You might discuss with other teachers at your school whether the SOLVE feature is something you want to introduce at all. There is no significant danger in familiarizing students with this calculator feature, as long as you require them to show sufficient work on activities. With that requirement, the worst that can happen is that they have another way to check their answers.

Find the Ferris Wheel: Graphing the two equations from Question 2b on the calculator at the same time is an easy way to verify predictions of the difference. You might use differing line thicknesses, as described earlier, to differentiate between the two curves.

Some Polar Practice: As polar coordinates are introduced, some students are bound to find the ▸**Polar** command (under the **CPX** menu on the MATH key). However, this calculator feature will not do their rectangular-to-polar coordinate conversion for them; it works only when converting complex numbers to polar form. The calculator does have a coordinate conversion feature, which students are less likely to locate, on the **ANGLE** menu. The Calculator Note "Converting Between Coordinate Systems" describes this feature; but, carefully consider the level of understanding your students are demonstrating before introducing this shortcut.

Pythagorean Trigonometry: As the screen below demonstrates, students need to be careful with parentheses when verifying the Pythagorean identity for Question 4. Students should develop the habit of using parentheses to make the desired order of operations clear, as shown in the first line. Note that the calculator will not accept $\mathbf{cos^2 45}$.

```
(cos(45))²
                .5
cos(45)²
                .5
cos(45²
     ⁻.7071067812
```

The TABLE and GRAPH features of the calculator provide a convenient way of vividly illustrating the Pythagorean identity. Begin by entering the function $\cos^2 x + \sin^2 x$ at the Y= screen in three pieces, as shown here. (**Y_1** and **Y_2** are entered by selecting them from the **Function** menu, found by pressing VARS and then using the right arrow to highlight **Y-VARS**.)

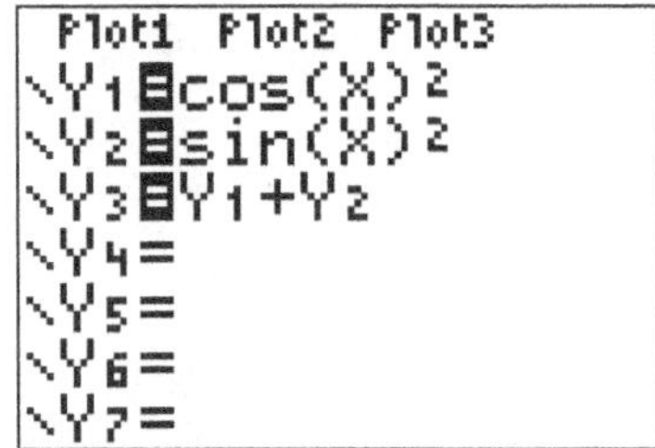

To use a table to verify the identity, set up the table by pressing 2ND [TBLSET] and then setting the variables to these values:

TblStart=0

ΔTbl=1

Indpnt: Auto

Depend: Auto

Press 2ND [TABLE] to view the table. The angles will be visible in the first (**X**) column and the values for $\cos^2 x$ and $\sin^2 x$ will be visible in the second (**Y_1**) and third (**Y_2**) columns, as in the first screen. Then use the right arrow to move the cursor to the fourth (**Y_3**) column, which is to the right of the columns that are initially visible. This will produce the second screen below. It is quite dramatic to see that $\cos^2 x$ and $\sin^2 x$, with their lengthy decimal values, always add to exactly 1.

X	Y1	Y2
0	1	0
1	.9997	3E-4
2	.99878	.00122
3	.99726	.00274
4	.99513	.00487
5	.9924	.0076
6	.98907	.01093

X=0

X	Y2	Y3
0	0	1
1	3E-4	1
2	.00122	1
3	.00274	1
4	.00487	1
5	.0076	1
6	.01093	1

Y3=1

To use a graph to further verify the identity, press WINDOW and set the variables to these values:

Xmin=-360

Xmax=360

Ymin=-2

Ymax=2

Press GRAPH to observe the graphs of the three functions being drawn, in the sequence in which they were entered at the Y= screen. Note that the sum of the two sinusoidal curves is always 1. Ask students to explain why both curves are drawn entirely above the x-axis, when they have learned previously that the sine and cosine curves are both centered about the x-axis.

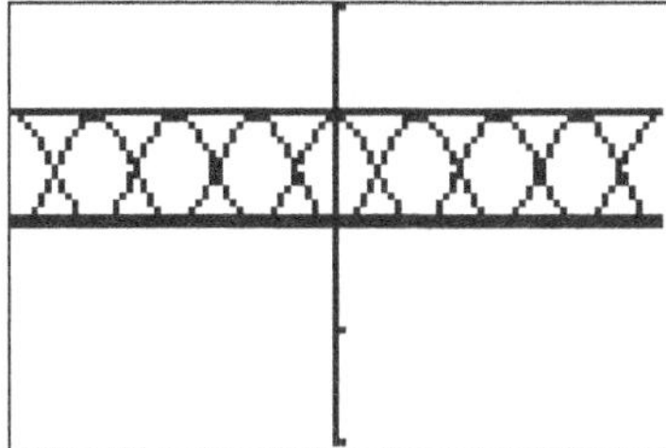

Coordinate Tangents: If students use the calculator to view the graph of the tangent function from Question 4 of, they are likely to be misled. The calculator may show some, but not necessarily all, of the asymptotes as solid lines. At the requested scale, it also truncates the tangent curve as it gets very close to the asymptote.

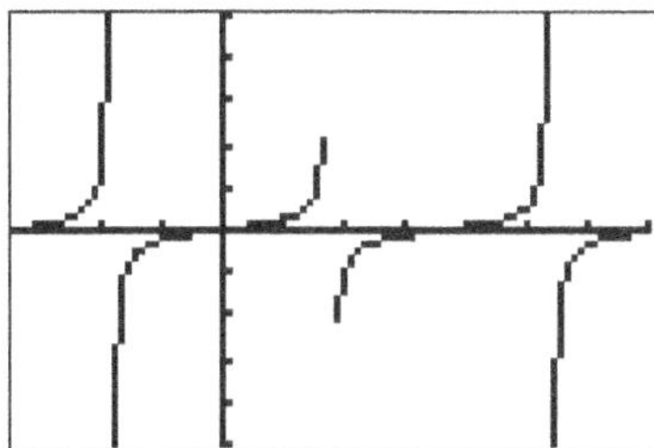

In discussing what happens to the tangent function as it approaches an asymptote, you can use the calculator table after entering the function into the Y= screen. Press 2ND [TBLSET] and set **Indpnt** to **Ask** and **Depend** to **Auto**, using the arrow and ENTER keys. (In this mode, the other settings will not matter.) Press 2ND [TABLE] and enter the angles at which you wish to evaluate the function.

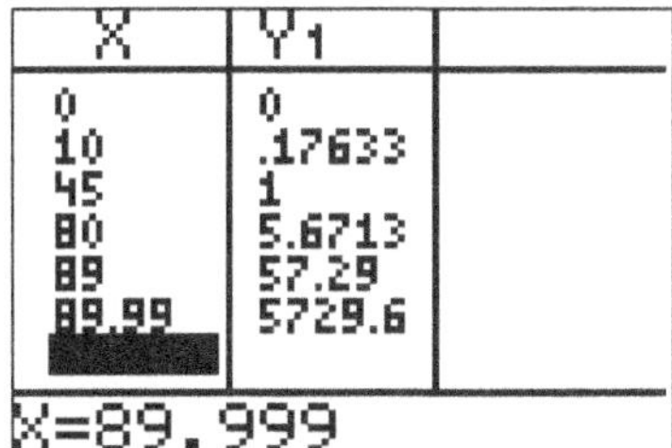

X	Y1	
0	0	
10	.17633	
45	1	
80	5.6713	
89	57.29	
89.99	5729.6	

X=89.999

Have students enter **tan 90** into their calculators and see what happens. They will get a "domain error" warning. Reinforce the terminology of *domain* and *range* by asking what the calculator is telling them in this error message.

Supplemental Activities:

A Polar Exploration: This activity asks students to investigate graphs of polar equations. This exploration can be accomplished very nicely on the graphing calculator. The instructions in the Calculator Note "Graphing Polar Equations" will help students get started.

Creating Tables on the Calculator

Throughout this unit, you will find it helpful to create In-Out tables as preparation for constructing graphs by hand. The calculator's TABLE feature can make this process much easier.

These instructions use as an example Question 1 of *Graphing the Ferris Wheel*. In that question, you graph the first 80 seconds of a Ferris wheel platform's movement, which is defined by the equation $h = 65 + 50 \sin 9t$.

1. Press Y= and enter the equation to be graphed. The calculator uses the variables **X** and **Y** in the Y= screen, so you must substitute these variables for t and h.

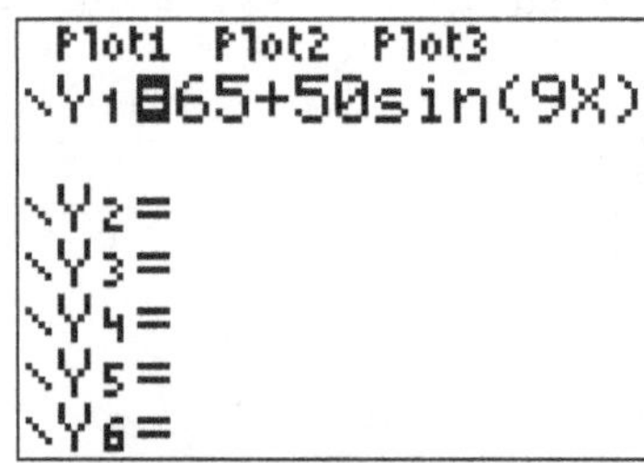

2. Because you are using a trigonometric function, press MODE and select **Degree**.
3. Press 2ND [TBLSET] to access the table setup screen.
4. Enter the first input value for your table at **TblStart**. Enter a value for **ΔTbl** (read "delta table"), which will be the size of the steps between entries in the input column of your table. For example, if you select **TblStart=20** and **ΔTbl=10**, the entries on the left side of your table will be 20, 30, 40, and so on. Select values for these variables that will yield a sufficient number of values for a smooth graph over the range specified in the activity. (The values 20 and 10 for **TblStart** and **ΔTbl** are not necessarily the best choice.)

5. Use the arrow and ENTER keys to highlight **Auto** for both **Indpnt** (the independent variable) and **Depend** (the dependent variable).
6. Press 2ND [TABLE] to view your table. Use the up and down arrows to scroll through the table to values that lie above or below the values visible on the screen. (The cursor must be in the **X** column to scroll upward.)

X	Y1	
20	65	
30	15	
40	65	
50	115	
60	65	
70	15	
80	65	

X=20

7. If you want to view only a few individually selected values, return to 2ND [TBLSET] and set **Indpnt** to **Ask**. Now press 2ND [TABLE] once more. Enter a value for **X** and press ENTER, and the corresponding value for **Y_1** will appear. The values for **TblStart** and **ΔTbl** will have no effect in this mode.

TABLE SETUP
TblStart=20
ΔTbl=10
Indpnt: Auto Ask
Depend: Auto Ask

Graphing Functions

The following instructions describe how to graph functions, using as an example the function developed for Question 1 of *Graphing the Ferris Wheel*. For that question, you graph the height of a particular Ferris wheel platform for the first 80 seconds of movement. That graph is defined by the equation $h = 65 + 50 \sin 9t$.

1. Press [Y=] and enter the equation to be graphed. Remember that the calculator uses the variables **X** and **Y** in the [Y=] screen, so you must first substitute these variables for *t* and *h*.

```
Plot1 Plot2 Plot3
\Y1=65+50sin(9X)
\Y2=
\Y3=
\Y4=
\Y5=
\Y6=
```

2. Because you are using a trigonometric function, press [MODE] and be sure to select **Degree**.

3. Press [WINDOW] to adjust the viewing window.

    a. Set **Xmin** and **Xmax** to the smallest and largest values of **X** in which you are interested. (These are defined by the question.) Choosing **Xmin** to be negative and **Xmax** to be positive will allow you to view the *x*-axis with your graph. Set **Ymin** and **Ymax** to include the smallest and largest values that you expect to see for $\mathbf{Y_1}$. This will require a little more thought. Don't be afraid to guess at the values; you can always make adjustments after viewing your graph. The values shown below are not your best choice.

```
WINDOW
 Xmin=-100
 Xmax=100
 Xscl=1
 Ymin=0
 Ymax=200
 Yscl=1
 Xres=1
```

    b. **Xscl** and **Yscl** determine the intervals at which the calculator places the tick marks along the axes. Because no numerical scale will appear beside these marks, it usually makes little difference what values you enter here.

    c. The setting, **Xres**, should be set to **1** to obtain maximum accuracy in your graph. Higher values, up to **8**, allow the graph to be drawn more quickly, but less accurately. (For example, a value of **5** for **Xres** would cause the calculator to evaluate and draw the function only at every fifth pixel.)

4. Press GRAPH to view your graph. If you find stray dots or lines on the screen or encounter a **DIM MISMATCH** error, you probably need to turn off STAT PLOT. This is most easily done by pressing 2ND [STAT PLOT] and selecting **4:PlotsOff**. Press ENTER when **PlotsOff** appears on your home screen.

Tracing Tips

Tracing a curve on a calculator graph to obtain coordinates with the desired degree of accuracy can be challenging. These instructions offer several suggestions, using Question 5 from *Sand Castles* as an example. In that question, you need to find the coordinates of the point that is on the function **Y1=20sin(29X)** and is one unit to the left of a point where **Y1** has a minimum.

1. Graph the function, with the window settings shown here.

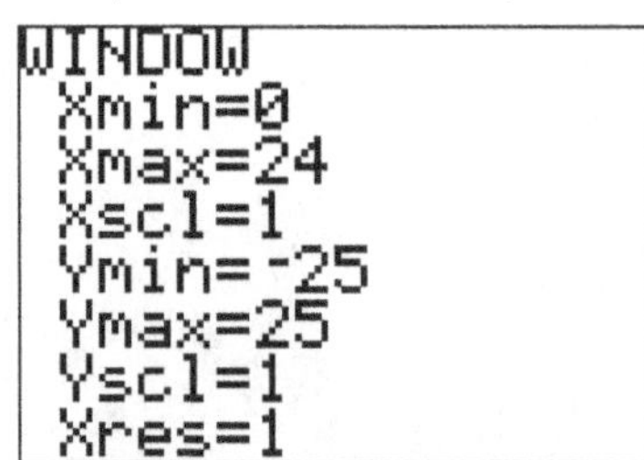

In Question 2, you found that a minimum occurs at $29t = 270$, because this makes sin $(29t)$ equal to -1. Solving that equation reveals that t is approximately 9.31 at the minimum. What you need to find, then, is where $t = 8.31$.

2. Press TRACE. The cursor will appear on the graph of your function, and coordinates for the cursor's location will appear at the bottom of the screen. (If the coordinates do not appear, press 2ND [FORMAT], use the down arrow to move the cursor to **CoordOn**, press ENTER to select it, and then press TRACE to return to your graph.)

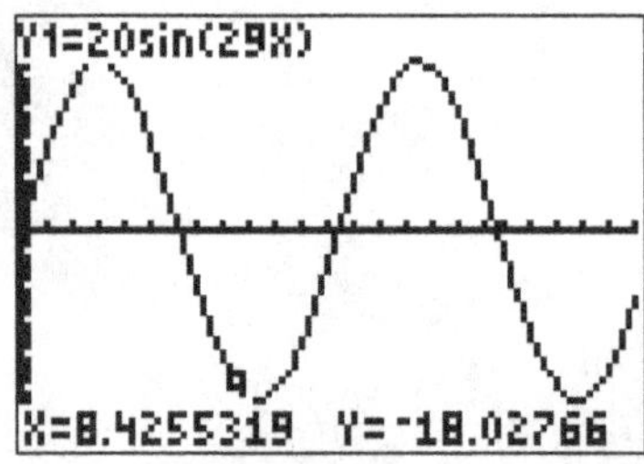

3. Using the left or right arrow key, try to find **x=8.31**. You will find that you can get close but cannot obtain exactly 8.31. Due to the size of the pixels, the values entered for the window range caused each pixel to represent a step of about 0.255. This is an awkward increment with which to work. You may have solved this difficulty in the past by repeatedly selecting **ZoomIn**.

4. However, there is a much easier way. Press the TRACE key, then press 8 . 3 1 ENTER. The cursor will move to that exact value, and the coordinates will be displayed.

5. Alternately, here is an approach that will give you a friendlier window within which to trace.

a. Press 2ND [QUIT] to return to the home screen.

b. Press . 1 STO> but do not press ENTER yet.

c. Press VARS to bring up the **VARS** menu. Press ENTER to select **1:Window**.

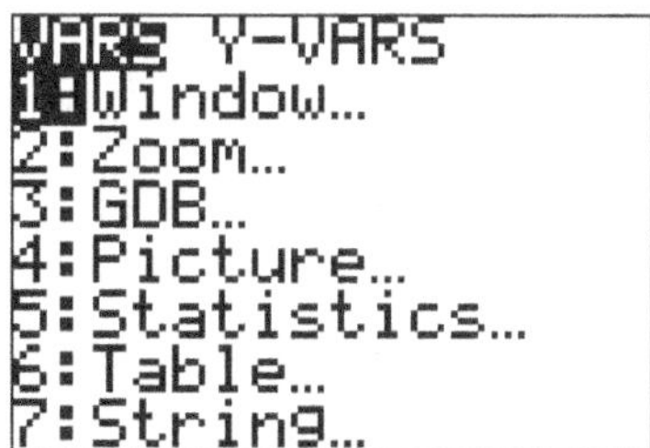

d. Select **8:ΔX** by pressing the 8 key or by scrolling down and highlighting it and then pressing ENTER.

e. Your home screen will now look like the one shown here. Press ENTER. You have just told the calculator to change the **Xmax** window setting such that each pixel represents an increment of exactly 0.1.

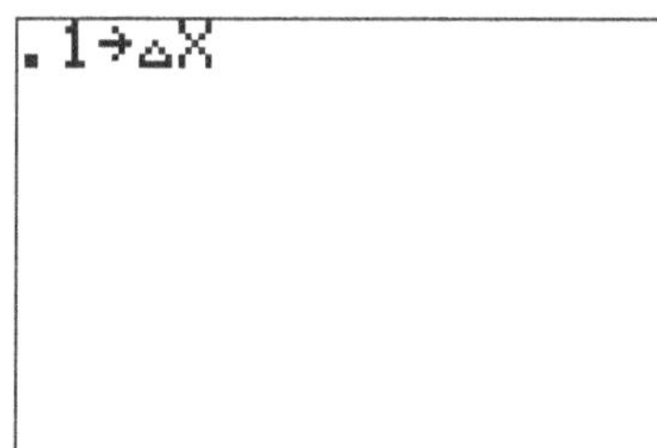

f. Press TRACE and use the left or right arrow key to locate **X=8.3**. Experiment with tracing to 8.31 instead of 8.3. You could do this by setting **Xmin** to 0.01 before changing **ΔX** to 0.1. You could also do it by changing **ΔX** to 0.01, but this will enlarge the graph so much that you will have to set **Xmin** much closer to 8.3 to be able to view the desired point.

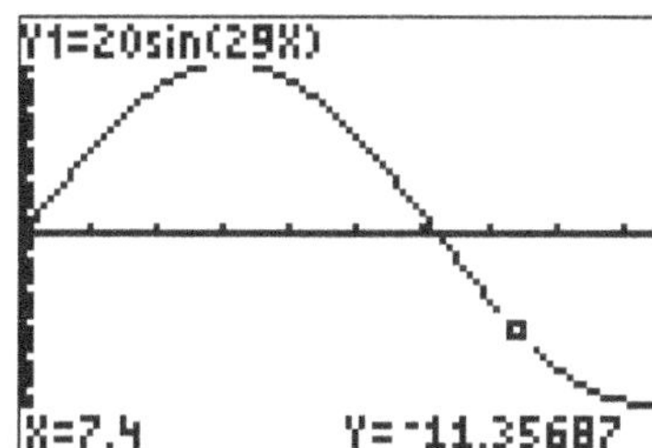

Solving "Sand Castles" with the CALC Menu

Sand Castles provides an excellent opportunity to learn to use some powerful calculator features to which you have not previously been exposed. Work through the suggestions below.

Question 1: Graphing the Function

Graph the function $w(t) = 20 \sin(29t)$ on your calculator. Set up the viewing window to match the one you used in drawing the graph by hand. Verify that the calculator graph looks like the one you drew by hand.

Question 2: Finding the Maximum and Minimum

To find the maximum value of the function, press 2ND [CALC]. Select **maximum** by pressing 4 or by highlighting it with the down arrow key and then pressing ENTER.

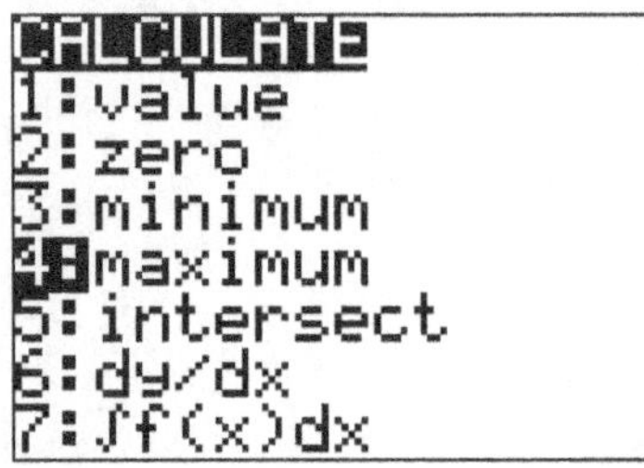

The calculator will prompt you to select the left boundary. Use the left or right arrow to move the cursor to any point on the curve to the left of one of the peaks, and press ENTER. The calculator will then prompt you to select a right boundary. Move the cursor to the right of the same peak and press ENTER. Finally, the calculator will ask you to make a guess. Move the cursor near the peak of the curve and press ENTER once more. In a moment, the calculator will display the coordinates of the maximum that you selected. Verify that this matches what you obtained from the equation.

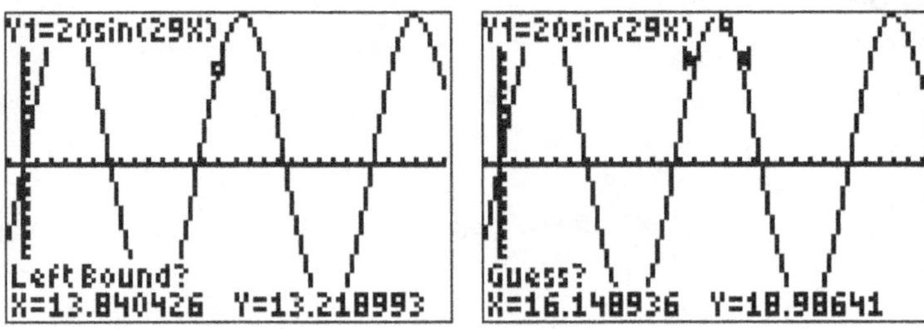

Use 2ND [CALC] to find the minimum in a similar manner.

Question 3: Finding the Zeros

To answer Question 3, you need to find the coordinates of two points where the function has a value of zero. These points need to be immediately to the left and right of a portion of the curve where the y-value is negative, so that the interval between them will represent the time in which the tide is below the average water line. Once again, the calculator provides a shortcut.

Go to the **CALCULATE** menu by pressing [2ND] [CALC] and select **2:zero**.

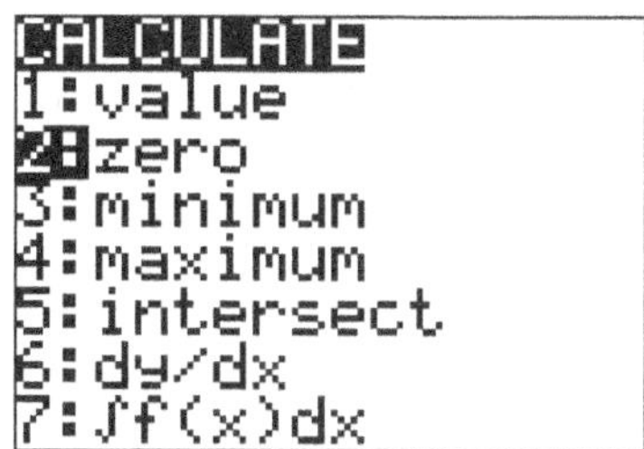

As before, you will be asked to select a left and a right boundary and to make a guess. Find the first zero by selecting boundaries on either side of the point where the curve first drops below the *x*-axis. Guess a value that is near the point of intersection with the axis.

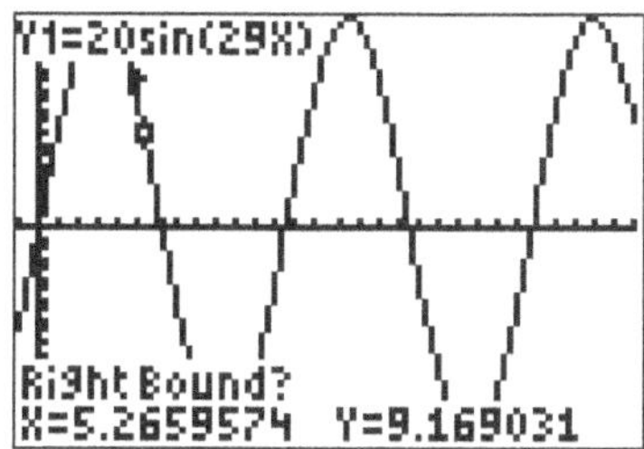

Find the second zero in a similar manner. Verify that these times agree with your calculations or estimations from Question 3.

Question 4: Finding Intersections

Question 4 requires that you find the time interval during which the function yields values less than −10. This will be the width of the shaded area in this illustration.

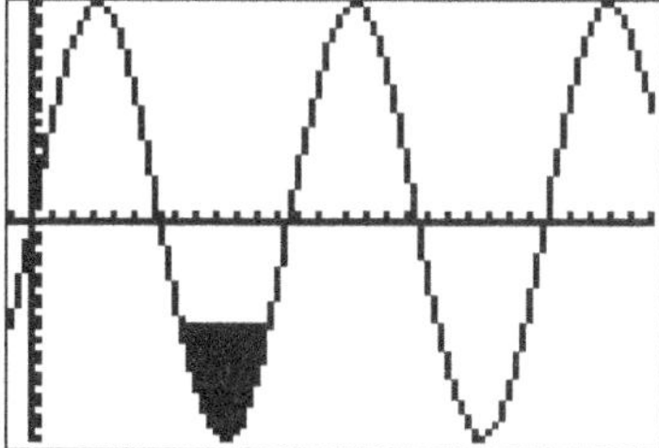

Return to the [Y=] screen and enter a second equation, **Y_2=-10**. Press [GRAPH], and both functions will be drawn on the same screen. The points you need to find are the two intersections of the curve and the line, which correspond to the left and right edges of the shaded area in the illustration above.

```
Plot1 Plot2 Plot3
\Y1=20sin(29X)
\Y2=-10
\Y3=
\Y4=
\Y5=
\Y6=
\Y7=
```

Press 2ND [CALC] and select **5:intersect**. The calculator will return to the graphing screen and will prompt you to select **First curve**. Your cursor will be located on the curve for **Y_1**. Press ENTER. You will then be prompted to select **Second curve**. Your cursor will be located on the line for **Y_2**. Press ENTER. Finally, you will be asked to make a guess. Place the cursor near one of the desired intersections and press ENTER again. The coordinates of the intersection will be displayed. Repeat this process to find the second intersection.

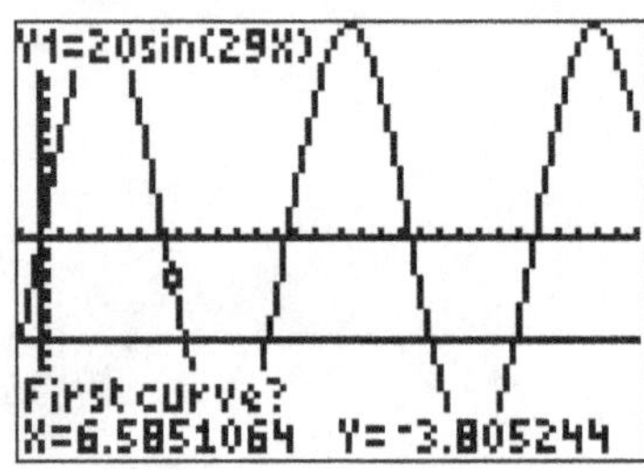

Guess-and-Check Without the Pain

In trying to solve *Moving Cart, Turning Ferris Wheel,* you have developed a complicated equation that defies algebraic solution. One way to solve it is through guess-and-check, but even that may be quite difficult. Placing parentheses within parentheses on the calculator quickly gets confusing, and having to reenter the entire equation for each new guess is tedious. The tips here will help you deal with these two difficulties.

Your equation probably includes the expression

$$\sqrt{\frac{57+50\sin 9W}{16}}$$

or something similar. These instructions use this expression as an example. You will need to expand what is explained here to include your entire equation.

1. Enter an initial guess for the value of your variable into your calculator's home screen. Use as many digits as needed to give the level of accuracy you want for your answer. Do not press [ENTER]. Instead, after your guess, press [STO>] [ALPHA] [W]. This stores your guess as the variable *W*. (You can use any variable to store information, but be aware that graphing a function will cause a value stored as *X* to change, so you should not use *X* here.)

```
18.00→W
```

2. Separate the command you entered in step 1 from the next one with a colon by pressing [ALPHA] [:].

3. Enter a portion of your function that does not require complicated parentheses and store it as variable A. Follow that with a colon. Continue to enter your function a piece at a time, without pressing the [ENTER] key, using variables to replace expressions as needed for simplicity. An example for the expression

$$\sqrt{\frac{57+50\sin 9W}{16}}$$

is shown here (with *B* representing the full expression); your function will be more complex. You may find it helpful to put your equation into a form in which all expressions containing the variable are on one side of the equation. In that way, you will only have to do a single calculation to check your guess.

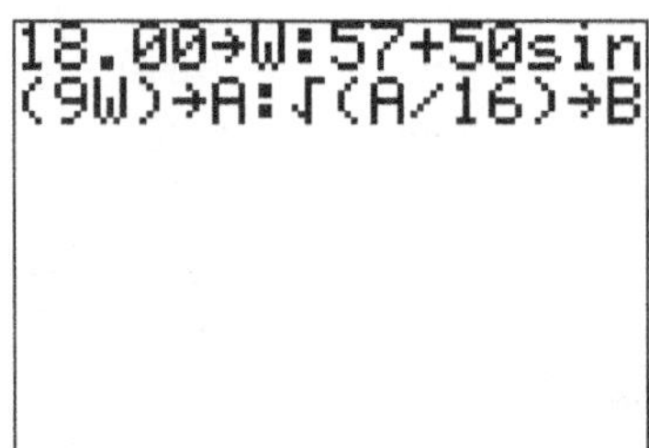

4. Press ENTER and check the result against the other side of your equation.

5. To adjust your guess, you do not need to reenter the entire function. Simply press 2ND [ENTRY] to recall the function to the screen, and then use the arrow keys to move to your guess. Edit the guess and press ENTER. Repeat as necessary.

```
18.00→W:57+50sin
(9W)→A:√(A/16)→B
      2.127951622
17.5■→W:57+50sin
(9W)→A:√(A/16)→B
```

Graphing a Complicated Function

The hardest part about graphing a complicated function is entering it into your calculator without order-of-operation mistakes. If you try to enter the function you developed for *Moving Cart, Turning Ferris Wheel,* you may find it difficult to determine where the many sets of parentheses that are needed should be opened and where they should be closed. These instructions use the expression

$$\sqrt{\frac{57+50\sin 9W}{16}}$$

as an example of how to do this without confusion. You will need to expand what is explained here to include your entire equation.

We could enter the expression as shown here, but notice how confusing the nested parentheses become. And this is only a part of the function you need to graph for *Moving Cart, Turning Ferris Wheel*. Fortunately, the calculator provides an easier way.

```
Plot1 Plot2 Plot3
\Y1=√((57+50sin(
9X))/16)
\Y2=
\Y3=
\Y4=
\Y5=
\Y6=
```

1. Press [Y=]. Choose a part of your function that can be entered without using more than one set of parentheses. If your function already contains parentheses, the expression inside of those parentheses is a good place to start. Enter this expression as **Y1**. In the example, we have started with the numerator of the expression within the radical symbol. Remember that the calculator requires that you substitute **X** for your variable.

   ```
   Plot1 Plot2 Plot3
   \Y1=57+50sin(9X)
   \Y2=
   \Y3=
   \Y4=
   \Y5=
   \Y6=
   ```

2. Enter another piece of your function as **Y2**. This can either be a completely separate piece or involve doing something more to the expression you entered as **Y1**. In our example, we can now simplify the original function by replacing **57+50sin(9X)** with **Y1**. The complicated expression

$$\sqrt{\frac{57 + 50 \sin 9W}{16}}$$

which was shown earlier as **√((57+50sin(9X))/16)**, can now be entered more simply as **√(Y_1/16)**.

To insert one function within the definition of another function, press VARS and use the right arrow to select the **Y-VARS** menu. From the **Y-VARS** menu, select **1:Function** by pressing the ENTER key. Finally, select **Y_1**, or whichever other variable you need.

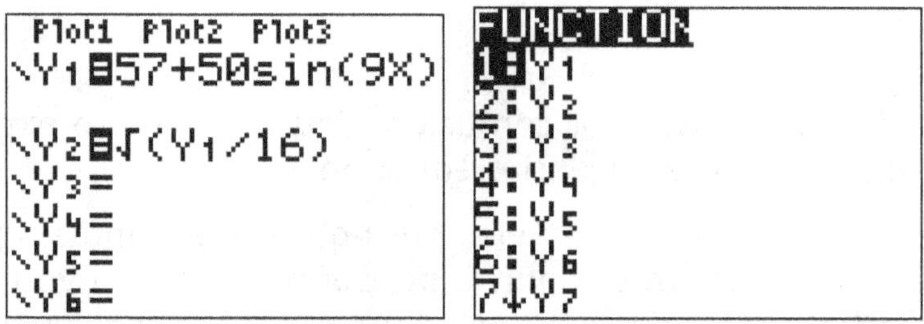

For a more complicated function, you may have to split it into three or four pieces, using **Y_3** and **Y_4**.

3. As the functions are defined now, pressing the GRAPH key would cause a curve to be drawn for each of the two functions we have used. But **Y_1** is merely a part of our original function; it is not something we want to graph. To tell the calculator not to graph **Y_1**, use the arrow keys to move the cursor onto the equal sign in front of **Y_1**. Press ENTER to remove the highlighting from that equal sign, making this function inactive with respect to being displayed on the graph. Repeat this for any other functions that you do not wish to see on your graph. Only those functions that have highlighted equal signs will be graphed.

```
Plot1 Plot2 Plot3
\Y1=57+50sin(9X)
\Y2=√(Y1/16)
\Y3=
\Y4=
\Y5=
\Y6=
```

Storing and Recalling a Graph Database

These instructions describe how to store and recall a graph data. This feature can help you to avoid the need to repeatedly enter complicated functions into your calculator by storing them for future recall. A single command will allow you to store or recall this information:

- All functions in the Y= screen and the display status of each
- The graphing mode
- All viewing window variables
- Format settings
- The line style for each Y= function

To store a graph database (GDB) press 2ND [DRAW], use the right arrow to display the **STO** menu, and press 3 to select **StoreGDB.** The **StoreGDB** command will be copied to the home screen. Press any number key to select one of ten variables to which this GDB can be stored, and then press ENTER.

```
StoreGDB 1
                Done
```

Use a similar procedure to recall the stored GDB, but select **RecallGDB** from the 2ND [DRAW] **STO** menu. Again, enter the variable number from which you wish to recall the GDB and press ENTER.

Caution: Recalling the GDB replaces all Y= functions. Any functions that are entered at the Y= screen will be erased when the GDB is recalled.

Using the Equation Solver

To use EQUATION SOLVER, your equation needs to be in a form with one side of the equation equal to zero. If your equation is complicated, as in *Moving Cart, Turning Ferris Wheel,* you will want to use the Y= screen first to break it up into smaller pieces. This process is described in the Calculator Note "Graphing a Complicated Function."

For an example, we will assume that you have broken your equation up and entered it into the Y= screen so that one side of your equation is represented by the variable **Y_3** and the other side by **Y_4**. So, your equation could now be expressed as **$Y_3=Y_4$**.

1. Manipulate your equation to create an equivalent equation in which one side of the equation is equal to zero. The equation **$Y_3=Y_4$** would become **$Y_3-Y_4=0$**. You will enter this equation into the calculator in a moment.

2. If you need to, press 2ND [QUIT] to return to the home screen. Press MATH. Use the down arrow to scroll all the way down to **0:Solver** and press ENTER.

3. You will enter your equation on the **EQUATION SOLVER** editing screen, which appears as shown here. If your screen does not say **EQUATION SOLVER** at the top, press the up arrow to get to this screen.

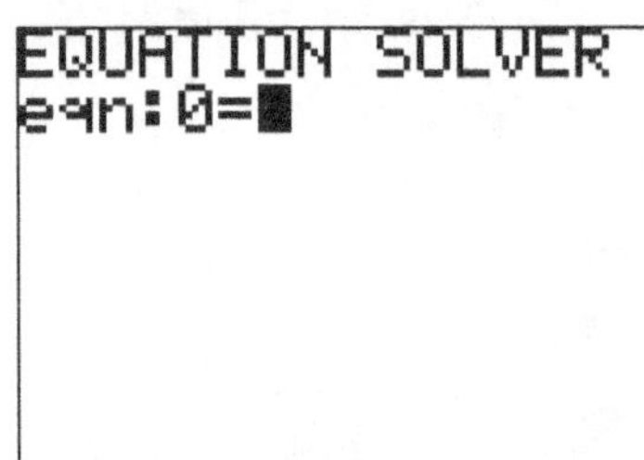

4. Enter your equation after the **0=** prompt. To enter **Y_3**, press VARS, use the right arrow to select the **Y-VARS** menu, select **1:Function** by pressing the ENTER key, and select **Y_3**. Do the same to select **Y_4**.

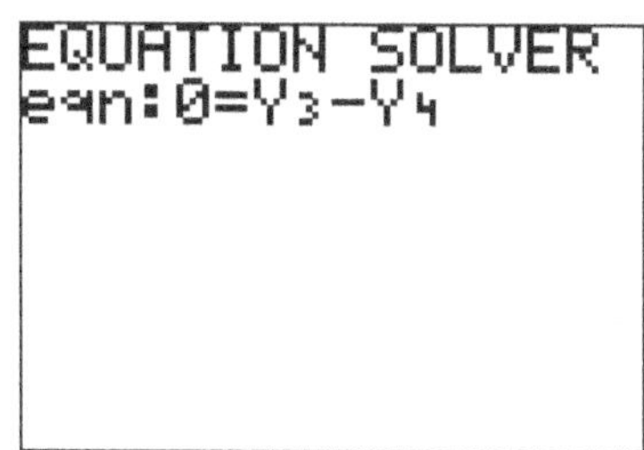

5. Press ENTER to obtain a screen similar to that shown here. The displayed value for the variable **X** will be whatever value was last stored under that variable, and it therefore has no significance.

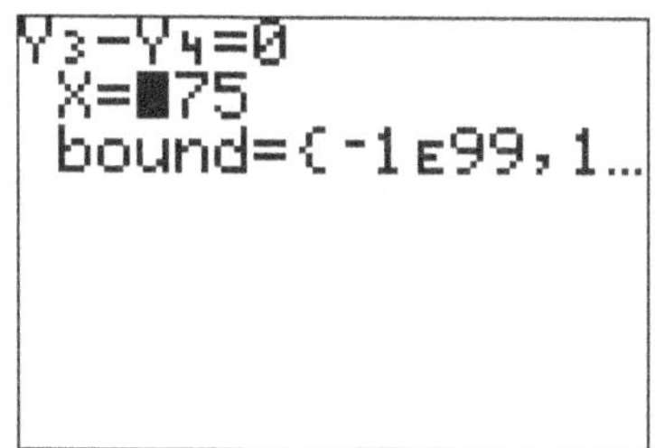

6. Change the displayed value of your variable to a reasonable guess of its actual value (the time at which the diver should be released from the Ferris wheel). Then use the down arrow to move to the boundaries that are displayed. The ... on the right side means there are more digits beyond the right edge of the screen, which can be viewed using the right arrow. These are the lower and upper boundaries, between which you want the calculator to find a value for your variable that makes your equation true. By default, the calculator chooses the smallest and largest numbers with which it can work as boundaries. Change these numbers to realistic lower and upper boundaries for your variable. For example, we know that X cannot be smaller than zero, because we are not interested in negative time. The two boundaries must be separated by a comma and enclosed in braces. Don't confuse braces ({,}) with brackets ([,]). Braces are found above the parentheses keys—press 2ND [{] and 2ND [}].

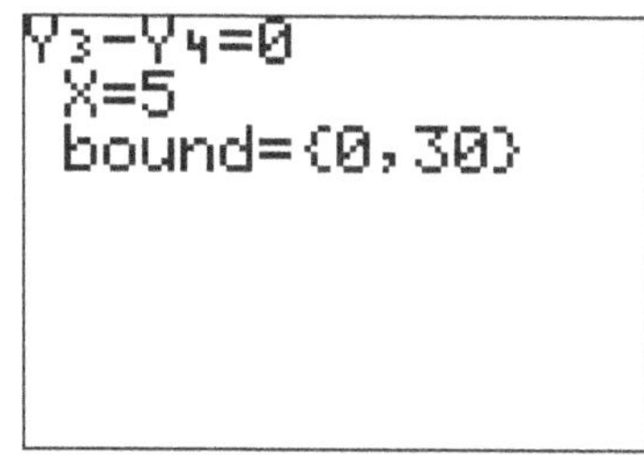

7. Use the up arrow to move your cursor back to the guess you made for your variable. Press ALPHA [SOLVE] (above the ENTER key); in a moment, the calculator will display the solution to **Y3-Y4=0**. The value displayed after **left-rt=** is the difference between the two sides of your equation, using the solution's value for the variable. If there is no round-off error, this should be zero.

Converting Between Coordinate Systems

The graphing calculator can perform conversions between rectangular and polar coordinates.

Converting from Rectangular to Polar Coordinates

1. Press MODE and make sure your calculator is set in the mode in which you wish to have angle measurements reported. This will usually be the **Degree** mode.
2. Press 2ND [ANGLE] (above the APPS key) and then press 5 to select **R▸Pr(**. The command will be copied to the home screen.

   Enter the rectangular coordinates, separated by a comma. Press ENTER. The calculator will display the *r*-value of the polar coordinates.

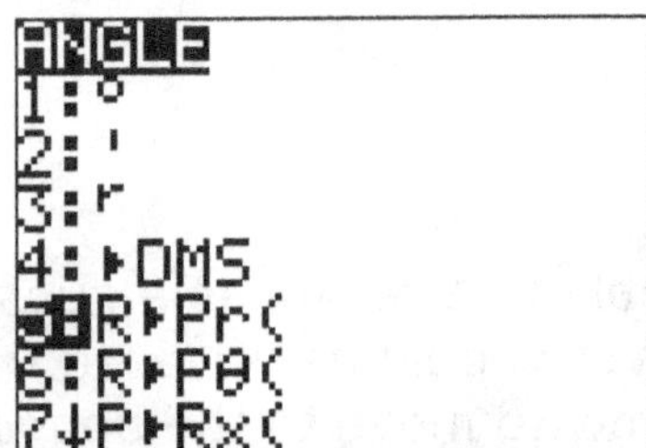

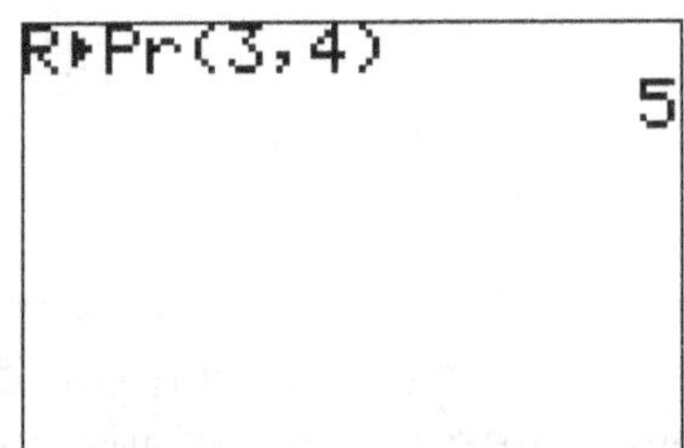

3. Press 2ND [ANGLE] again and then press 6 to select **R▸P*θ*(**. The command will be copied to the home screen.

   Enter the rectangular coordinates again and press ENTER. The calculator will display the *θ*-value of the polar coordinates. This screen shows a conversion of the rectangular coordinates (3, 4) to the polar coordinates (5, 53.13).

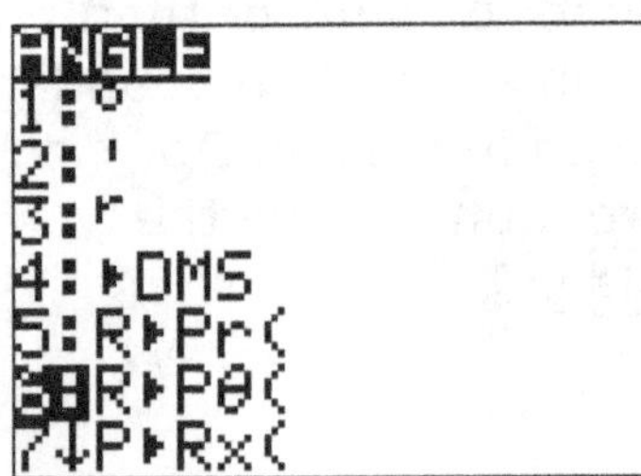

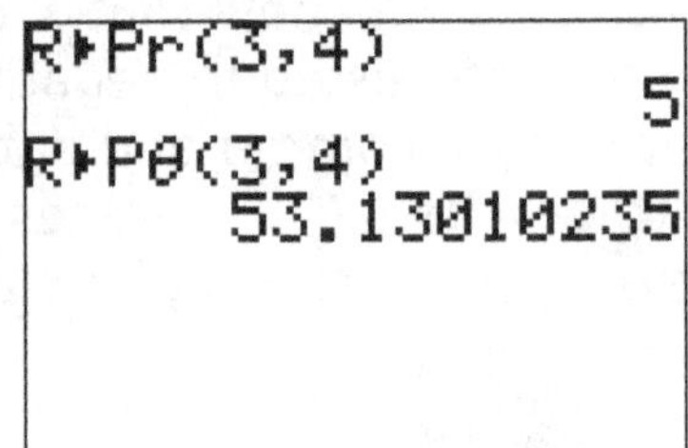

Converting from Polar to Rectangular Coordinates

1. Press MODE and make sure your calculator is set in the mode in which you have the angle measurement. This will usually be the **Degree** mode.
2. Press 2ND [ANGLE] (above the APPS key) and then press 7 to select **P▸Rx(**. The command will be copied to the home screen.

   Enter the polar coordinates, separated by a comma, with the *r*-value before the *θ*-value. Press ENTER. The calculator will display the *x*-value of the rectangular coordinates.

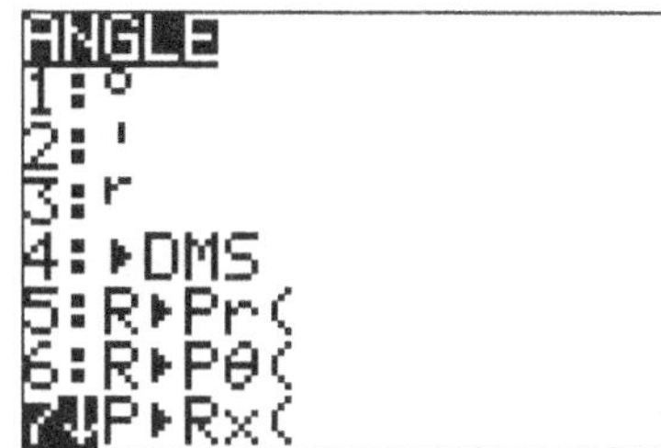

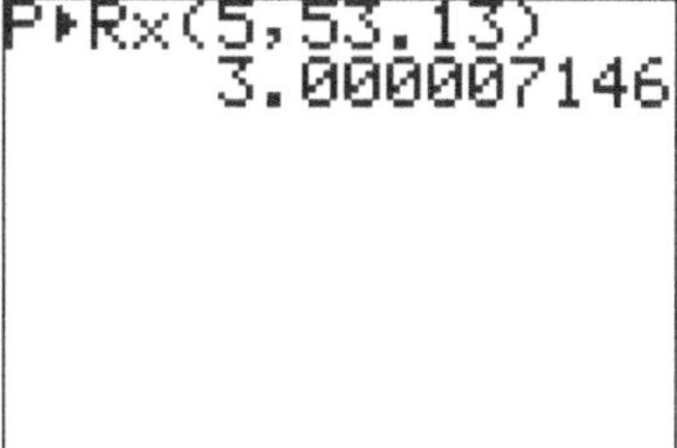

3. Press 2ND [ANGLE] again and then press 8 to select **P▶Ry(**. (You can view this command by using the down arrow to scroll downward beyond the last menu option visible on the screen.) The command will be copied to the home screen.

   Enter the polar coordinates again and press ENTER. The calculator will display the *y*-value of the rectangular coordinates. This screen shows a conversion from polar coordinates of (5, 53.13) to rectangular coordinates of approximately (3, 4).

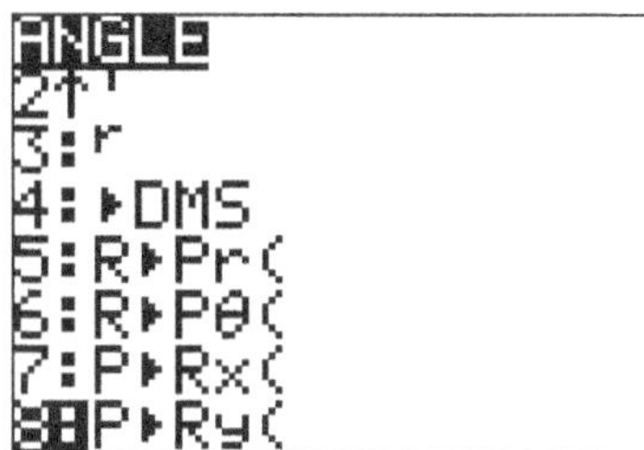

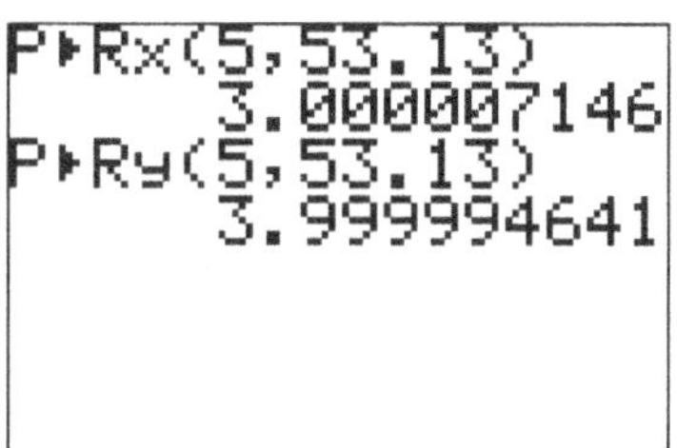

Graphing Polar Equations

These instructions explain how to graph polar equations on the graphing calculator.

1. Press MODE, use the arrow keys to highlight **Pol** and press ENTER. This shifts the calculator into polar mode. While you are at the MODE screen, be sure that **Degree** is highlighted.

2. Press Y= and enter your equation. The equation must first be in a format with *r* alone on one side. Use the X,T,θ,n key to enter **θ**.

```
 Plot1 Plot2 Plot3
\r1=sin(θ/2)
\r2=
\r3=
\r4=
\r5=
\r6=
```

3. Press WINDOW and enter appropriate values for the various variables. In most cases, you will want to graph from **θmin=0** to **θmax=360**. The variable **θstep** controls how finely the data items for the graph are calculated. Setting **θstep=1** will cause the calculator to determine the coordinates at every integer number of degrees. A larger number will draw more quickly but will also yield a coarser approximation of the graph.

```
WINDOW
 θmin=0
 θmax=360
 θstep=1
 Xmin=-1
 Xmax=1
 Xscl=.5
↓Ymin=-1
```

Notice that even though this will be a polar graph, the window is still dimensioned in terms of *x* and *y*. While it may be difficult to predict what your graph will look like, it is usually fairly simple to at least estimate the greatest possible value of *r* by using what you know about trigonometric functions. For example, in the equation shown in the preceding illustration, we know that the sine always ranges between −1 and 1. A safe window to use would extend one unit in each direction, so **Xmin** and **Ymin** are set to −1 and **Xmax** and **Ymax** are set to 1. **Xscl** and **Yscl** simply control the distance between tick marks on the axes.

4. Because we are using a coordinate system based on the circle, the graph will be distorted if the window is not proportioned correctly. That is, circles will look more like ovals. The easiest way to avoid this is to use **Zoom Square** after setting up the window variables. This feature will enlarge your window in one direction to make each step in the horizontal direction equal in size to a step in the vertical direction. Press [ZOOM] and then [5] to select **5:ZSquare**. The graph will be drawn.

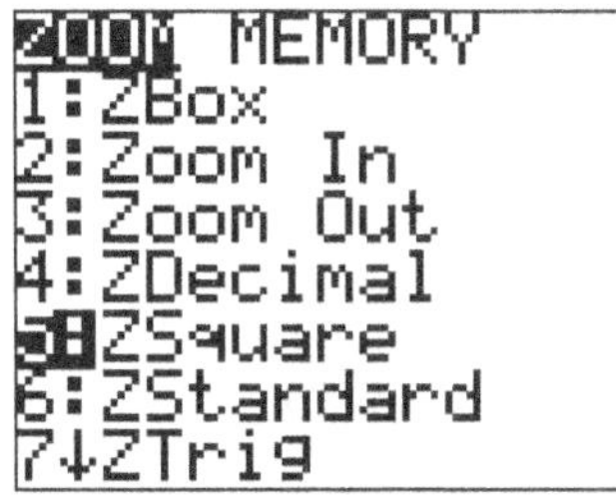

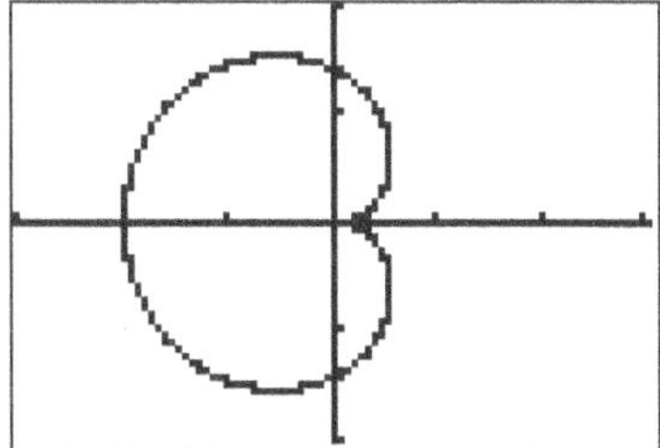

5. If you wish to trace the function using polar coordinates, you must select polar grid coordinates in the window format menu. To do this, press [2ND] [FORMAT], highlight **PolarGC**, and press [ENTER]. Press [TRACE] and use the left and right arrow keys to trace the function.

www.ingramcontent.com/pod-product-compliance
Lightning Source LLC
Chambersburg PA
CBHW051345200326
41521CB00014B/2489

* 9 7 8 1 6 0 4 4 0 1 1 6 5 *